LATEST FINDINGS OF OMEGA-3 LONG CHAIN-POLYUNSATURATED FATTY ACIDS: FROM MOLECULAR MECHANISMS TO NEW APPLICATIONS IN HEALTH AND DISEASES

Editor: Maricela Rodriguez-Cruz; M. Sc., PhD

Molecular Biology Laboratory , Unit of Research in Medical Nutrition, Pediatric Hospital, "Centro Médico Nacional Siglo XXI", Mexican Institute of Social Security, Mexico City, México.

Co-Editor: Mardia López-Alarcón; MD, PhD

Unit of Research in Medical Nutrition, Pediatric Hospital, "Centro Médico Nacional Siglo XXI", Mexican Institute of Social Security, Mexico City, México.

CONTENTS

FOREWORD

The growing interest on this group of these long-chain polyunsaturated fatty acids derived from the alpha linolenic acid known generically as omega-3 fatty acids is due to their ubiquity in many bodily functions both, in health and disease. Those functions include its role as integral part of cell membranes and its diverse role as biochemical signal involved in the neural development, conductivity of the electric system of the heart, coagulation, chronic inflammation involving asthma, cardiovascular risk and certain type of cancers and modulating gene expression among others. Omega-3 fatty acids are of particular interest for developing countries, because the typical dietary intake is inadequately low in many of those countries. At the same time their burden of diseases include a high proportion of cardiovascular diseases, asthma and other linked to low intensity inflammation, that may be alleviated by the anti-inflammatory capacity of an adequate nutritional status of Omega-3 fatty and a good balance with its pro-inflammatory Omega-6 counterparts.

The editors of this book have made a very comprehensive effort to provide readers with a state-of-the art review of many of those topics, gathering an excellent roster of international experts, including the experience of their research group. This Mexican research group has an international reputation because of their seminal contributions, among others, the demonstration that elongases and desaturases were present in the mammary gland, inferring that down-stream Omega-3 fatty acids in the milk are synthesized by the gland.

I had the privilege to witness the evolving career of both editors, Maricela Rodriguez-Cruz and Mardia López-Alarcon from the beginning until seeing them transformed into mature scientists. This book demonstrates their professional maturity and it is for me, an opportunity to voice out my personal pride for the opportunity I had to walk with them by the paths of science.

This book offering to readers interested in biochemical or medical aspects of Omega-3 fatty acids recent and relevant information from dietary recommendations to identified dietary sources; the impact of dietary interventions on specific health problems and biochemical details for their synthetic pathways will be of great help in better understanding of their functions and their role in health.

Salvador Villalpando M.D.; Ph.D.

PREFACE

The knowledge generated about the benefits of the n-3 long chain polyunsaturated fatty acids (LC-PUFAs) is huge and relevant to different medical areas such as endocrinology, cardiology, pediatrics, nutrition, immunology, psychiatry and oncology. Therefore, with this book, the massive medical and scientific community will have rapid access to the related advances to their field. Thus, this book has been prepared to give details about the state of the art regarding the n-3 LC-PUFAs in nutrition and medicine, in physiology and gene regulation, their food sources and recommendations, as well as the likely protective mechanisms of these fatty acids in major diseases.

There is considerable interest in the impact of n-3 LC-PUFAs in mitigating the morbidity and mortality caused by chronic diseases. Increased intakes of n-3 LC-PUFAs such as eicosapentaenoic (EPA) and docosahexaenoic (DHA) acid are related with positive effects on cell and tissue function and human health. It has been demonstrated that supplementation with these fatty acids decreases blood triglycerides; slows the buildup of atherosclerotic plaques; lowers blood pressure slightly; reduces the risk of death, heart attack, and arrhythmias; improves neurological and cognitive development in newborns, decreases insulin resistance, and leads to reduction of hepatic steatosis. Beneficial effects have been observed also in human models of diabetes, obesity, and cardiovascular diseases, and lately, increased evidence about positive effects of n-3 LC-PUFAs on depression and its common comorbid physical conditions has been released. In addition, our research group has demonstrated recently that DHA administered in the acute phase of infection with sepsis, modulates cytokines production, and decreases the effect of infection on appetite protecting the nutritional status of children. Most of the health benefits of n-3 LC-PUFAs can be attributed to their influence of the physical nature of cell membranes and membrane protein-mediated responses, eicosanoid generation, cell signaling and on expression of genes involved in cellular key processes.

As it is known, n-3 LC-PUFAs are not efficiently synthesized in humans. Therefore, they need to be obtained directly from the diet. Thus, it is important to know the main sources of n-3 LC-PUFAs to promote the consumption of these fatty acids. Accordingly, based in epidemiological and clinical studies it is necessary to spread the knowledge about the recommendations of LC-PUFAs of different international health organizations such as FAO/WHO.

Authors of this book are well-recognized researchers in their specific field. All of them have made important contribution to the knowledge of the beneficial effects of n-3 LC-PUFAs at different levels. We extensively thank them for the time dedicated to the writing of this book, which we are sure will help to understand the mechanisms used by such fatty acids and their application in translational medicine.

There are some people who have helped us with the content of this book. Specifically, we thank Engineer **Salomón Alarcón Fuentes** by his contribution in the design of the cover. We also warmly thank our husbands and my mother for the time we stole from them for the preparation of this book.

Maricela Rodriguez-Cruz, M. Sc, Ph. D.
Mexican Institute of Social Security
México

&

Mardia López-Alarcón, M. D., Ph. D.
Mexican Institute of Social Security
México

CONTRIBUTORS

Chih-Chiang Chiu M. D.

Department of Psychiatry, Taipei City Psychiatric Center
Taipei City Hospital, Taipei,
Taiwan.
Department of Psychiatry, School of Medicine
Taipei Medical University, Taipei
Taiwan.
King's College London (Institute of Psychiatry)
London
United Kingdom.

Blanca Estela Del Río Navarro M. D.

Department of Allergy and Clinical Immunology
Hospital Infantil de México "Federico Gómez" México D.F.

Dino Roberto Pietropaolo Cienfuegos M. D.

Department of Allergy and Clinical Immunology
Hospital Infantil de México "Federico Gómez" México D.F.

Jane Pei-Chen Chang, M.D.

Department of Psychiatry & Mind-Body Interface Laboratory (MBI-Lab)
China Medical University Hospital, Taichung
Taiwan.
Graduate Institute of Clinical Medical Science
China Medical University Hospital, Taichung
Taiwan.

Javier Mancilla-Ramírez

Instituto Nacional de Perinatología "Isidro Espinoza de los Reyes"
Mexico City
Mexico.

Kuan-Pin Su, M.D., Ph.D.

Department of Psychiatry & Mind-Body Interface Laboratory (MBI-Lab)
China Medical University Hospital
Taichung
Taiwan.
Graduate Institute of Neural and Cognitive Sciences
China Medical University, Taichung
Taiwan.

Marco Antonio Góngora Meléndez, MD.

Department of Allergy and Clinical Immunology
Hospital Infantil de México "Federico Gómez" México D.F.

María Teresa Villarreal-Molina Ph. D.

Unit of Molecular Biology and Genomic Medicine
Instituto Nacional de Ciencias Médicas y Nutrición Salvador Zubirán. México, D. F.

Mariela Bernabe-Garcia, M. Sc.
Unidad de Investigación Médica en Nutrición
Hospital de Pediatría, C.M.N. Siglo XXI
Instituto Mexicano del Seguro Social. México, D. F.

Robert J Stewart M. D.
King's College London (Institute of Psychiatry)
London
United Kingdom.

Samuel Canizales-Quinteros M. Sc., Ph. D.
Unit of Molecular Biology and Genomic Medicine
Instituto Nacional de Ciencias Médicas y Nutrición Salvador Zubirán. México, D. F. Departamento de Biología,
Facultad de Química, UNAM.

Shih-Yi Huang Ph. D.
School of Nutrition and Health Sciences
Taipei Medical University, Taipei
Taiwan.

CHAPTER 1

Molecular Mechanisms for the Synthesis and Genetic Regulation of Long-Chain Polyunsaturated Fatty Acids

Maricela Rodríguez-Cruz, M. Sc., Ph.D.

Laboratorio de Biología Molecular, Unidad de Investigación Médica en Nutrición, Hospital de Pediatría, Centro Medico Nacional Siglo XXI, IMSS, México City, México.

Abstract: Essential fatty acids—linoleic and linolenic—are metabolized to long-chain polyunsaturated fatty acids (LC-PUFAs) such as arachidonic acid (AA) and eicosapentaenoic acid (EPA) and docosahexaenoic acid (DHA), respectively, throughout a series of desaturation and elongation steps. LC-PUFAs are essential for a variety of physiological functions including brain development, cardiac function, inflammatory response, and homeostasis. These roles at the cellular level include modulation of signal transduction via effect of bioactive cell membranes and by regulating the expression of a wide array of genes through different transcription factors such as peroxisome proliferator-activated receptors (PPARs), sterol regulatory-element binding protein (SREBP), carbohydrate response-element binding protein (ChREBP) and nuclear factor κB (NFκB) mainly to control the transcription of target genes including those encoding proteins involved with lipid and carbohydrate metabolism, thermogenesis, and cell differentiation. However, more work is required to delineate these actions and to have a better understanding of the beneficial role of LC-PUFAs in order to comprehend the action of these fatty acids in the pathogenesis of various diseases. Integrative analysis including nutritional, biochemical, genetic and immunological studies may provide information about the identification of specific molecular mechanisms involved in the beneficial effects of n3 LC-PUFAs such as DHA and EPA intake and their metabolic derivates on health promotion or disease burden.

1. INTRODUCTION

Dietary fat may influence a variety of physiological events in the human body and may have an impact on the pathogenesis of various diseases. Properties of fat are influenced by fatty acids components. Fatty acids are classified as either saturated or unsaturated, respectively, depending on absence or presence of a carbon-to-carbon double bond. Unsaturated fatty acids are further divided into two subgroups: monounsaturated fatty acids containing only one double bond and polyunsaturated fatty acids (PUFAs), which harbor two or more double bonds [1]. PUFAs are metabolized to long-chain PUFAs (LC-PUFAs) such as docosahexaenoic acid (DHA), eicosapentaenoic acid (EPA) and arachidonic acid (AA) via a common fatty acids desaturase/elongase system [2].

Primarily through dietary studies, LC-PUFAs such as DHA have been positively linked to an enormous variety of human diseases including cancer, heart disease, rheumatoid arthritis, asthma, lupus, alcoholism, visual acuity, kidney disease, respiratory disease, peroxisomal disorders, dermatitis, psoriasis, cystic fibrosis, schizophrenia, depression, dyslexia, neurological and brain development, malaria, multiple sclerosis and even migraine headaches. In order for one simple molecule to affect so many seemingly unrelated processes, it must function at a fundamental level common to most cells [3,4]. The mechanisms implicated in these processes are complex and multiple, reflecting the extraordinary diversity of the functions exercised by the PUFAs from the modulation of dynamic properties of the membranes to the production of active mediators and regulation of gene expression. As such, considerable clinical and basic scientific research has been directed at understanding the biochemical and molecular basis of the effects of fatty acids on complex physiological systems impacting human health.

Our understanding of the role that dietary LC-PUFAs play in these chronic diseases is complicated by the fact that those fatty acids have many physiological roles. This review represents an overview regarding the molecular mechanism of LC-PUFAs (AA, EPA and DHA) triggered to maintain cell homeostasis. With the recent information, the aim is to provide a better understanding about the beneficial role of dietary LC-PUFAs to comprehend the action of these fatty acids in the pathogenesis of various diseases.

*Address correspondence to Dr. Maricela Rodríguez-Cruz: Apartado postal C-029 C. S.P.I. "Coahuila" Coahuila No. 5, Col. Roma 06703. México, D. F., México. Tel. +52 56276900, ext. 22483, 22484. Fax +52 56276944; E-mail: maricela.rodriguez.cruz@gmail.com

POLYUNSATURATED FATTY ACIDS

Essential Fatty Acids

PUFAs are hydrocarbon chains with two or more double bonds situated along the length of the carbon chain. Depending on the location of the first double bond relative to the methyl terminus of the molecule, they may be classified as either omega or n (ω or n) -6 or -3. In mammals, the fatty acids linoleic acid (LA n6) and alpha linolenic acid (LNA n3) are essential fatty acids (EFAs) and cannot be endogenously synthesized by mammals. Both should be provided in the diet [5]. In westernized diets, LA is the primary PUFA followed by LNA. Once consumed, these fatty acids are further metabolized within mammalian cells. The requirement for EFAs is due to the absence in mammals of fatty acyl-CoA desaturases that introduce a double bond between position 9 of a monounsaturated fatty acids and the methylene end of the hydrocarbon chain. Both EFAs can be further metabolized to LC-PUFAs through a series of desaturation and elongation steps. LA is metabolized to AA, whereas LNA can be metabolized to EPA and ultimately to DHA [5,6].

Synthesis of LC-PUFAs

Once consumed, EFAs are further metabolized within mammalian cells. Synthesis of LC-PUFAs involves a first insertion of a double bond at the $\Delta 6$ position of 18:3 cis 9,12,15 n3 (LNA) or 18:2 cis 9,12 n6 (LA) by delta 6 desaturase ($\Delta 6$D) to synthesize stearidonic acid (18:4 n3) and gamma-linolenic acid (18:3 n6), respectively (Fig. 1). This is followed by an elongation step and then a second insertion of a double bond at the $\Delta 5$ position of the fatty acid by delta 5 desaturase ($\Delta 5$D) to form the key intermediates 20:4 cis 5,8,11,14 n6 (AA) and 20:5 cis 5,8,11,14,17 n3 (EPA). Then, 20:4 n6 and 20:5 n3 are further metabolized to docosapentaenoic acid (22:5 n6) and 22:6 cis 4,7,10,13,16,19 n3 (DHA), respectively, by a third elongation followed by insertion of a double bond into the $\Delta 6$ position of the 24-carbon substrate by $\Delta 6$D. [2]. First and second $\Delta 6$ desaturation steps reportedly can be catalyzed by the same $\Delta 6$D enzyme [7]. Both pathway syntheses of LC-PUFAs use the same series of enzymes; therefore, a competition exists between the n3 and n6 fatty acids families for metabolism with an excess of one causing a significant decrease in the conversion of the other. AA, EPA and DHA are converted to eicosanoids by the cyclooxygenase and lipoxygenase enzymatic pathways. Eicosanoids are involved in pain, inflammation, immune function, vasoconstriction, vasodilation and platelet aggregation [2]. Availability of 20- and 22-carbon n6 and n3 polyenoic fatty acids is determined by activity of $\Delta 5$D and $\Delta 6$D desaturases that catalyze the introduction of a *cis* double bond at the C5 and C6 positions, respectively, with strict regioselectivity and stereoselectivity. In addition, a four-step reaction cycle of the fatty acid chain elongation system is required to add two carbons on the fatty acyl-CoA intermediates derived from the desaturation step. Members of the elongation-of-very-long-chain-fatty acids (Elovl) gene family encoding fatty acids elongases are believed to perform the first regulatory condensation step in this elongation cycle [8]. Two fatty acids elongases, Elovl5 and Elovl2, have been identified and characterized as important enzymes in the biosynthesis of LC-PUFAs in mammals [9]. Functional studies on these enzymes suggest that Elovl5 is mainly involved in the elongation of the C18–C20 PUFAs, whereas Elovl2 has been shown to elongate C20–C22 PUFAs [10].

Desaturation of a fatty acid first involves the enzymatic removal of hydrogen from a methylene group in an acyl chain, a highly energy-demanding step (NADH or NADPH) that requires one activated oxygen intermediate. $\Delta 5$D and $\Delta 6$D are microsomal enzymes that are thought to be a component of a three-enzyme system that includes NADH-cytochrome *b5* reductase, cytochrome *b5*, and a cyanide-sensitive desaturase containing nonheme iron [11]. Most known desaturase genes involved in LC-PUFAs biosynthesis contain the N-terminal cytochrome *b5* domain (HPGG) as electron donor and three histidine box motifs ("HXXXH, HXXHH and QXXHH") conserved from human to microalga [12,13]. Fatty acids elongation occurs in the endoplasmic reticulum (ER) and microsomal fatty acids elongation adds two-carbon units to pre-existing fatty acyl-CoAs using malonyl-CoA as the donor and NADPH as the reducing agent. Fatty acids elongation requires four sequential reactions performed by individual proteins, which may be physically associated. Three of the four enzymatic activities are localized to the cytoplasmic side of ER membranes, whereas the enzyme performing the third step is suggested to be embedded in the membrane [14]: 1) condensation between fatty acyl-CoA and malonyl-CoA to generate 3-ketoacyl-CoA, 2) reduction of 3-ketoacyl-CoA using NADPH to form 3-hydroxyacyl-CoA, 3) dehydration of 3-hydroxyacyl-CoA to *trans*-2-enoyl-CoA, and 4) reduction of *trans*-2-enoyl-CoA to fully elongated acyl-CoA. Each reaction is catalyzed by different enzymes encoded by separate genes. Elovls are responsible for the rate-controlling and initial condensation reaction

[15,16]. The rate of fatty acids elongation is determined by the activity of the elongase (condensing enzyme) and not the reductase or the dehydratase [16]. It has been demonstrated that condensing enzymes alone is sufficient for specific induction of LC-PUFAs synthesis; thus, it is the limiting factor for this synthesis [14].

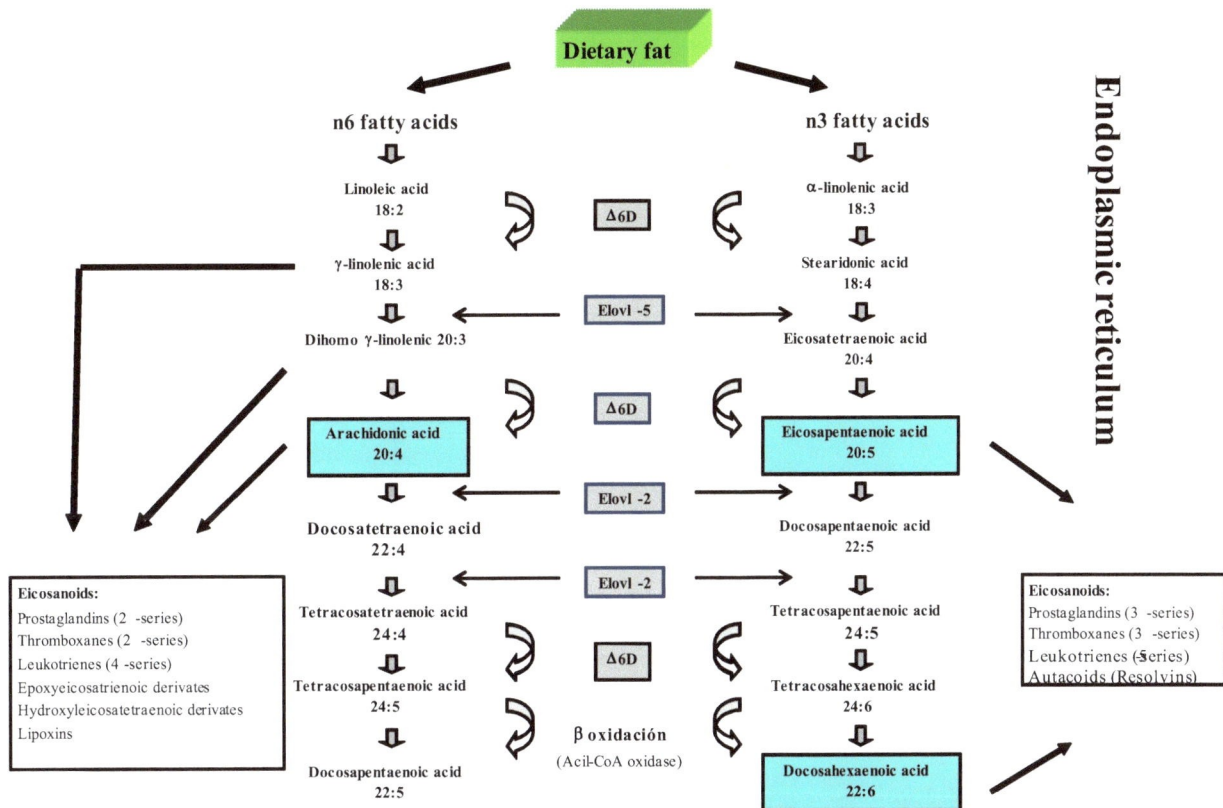

Figure 1: Synthesis of long-chain polyunsaturated fatty acids.

MOLECULAR MECHANISMS OF LC-PUFAs

LC-PUFAs such as AA and DHA are essential for a variety of physiological functions including brain development, cardiac function, inflammatory response, and homeostasis [17,18]. These functional roles of LC-PUFAs at the cellular level include membrane lipids such as ligands for nuclear receptors and transcription factors in the nucleus to regulate the transcription of a variety of genes, precursors of signalling molecules such as eicosanoids, and through direct interactions with proteins by multiple and complex implicated mechanisms reflecting the extraordinary diversity of the functions exercised by the LC-PUFAs, ranging from the modulation of dynamic properties of the membranes to the production of active mediators and regulation of gene expression [19].

Thus, regulation of cell functions by LC-PUFAs may occur on two general levels: modulation of signal transduction via effect of bioactivity on cell membranes and regulation of gene transcription.

Effect of Bioactivity of LC-PUFAs on Cell Membranes

A multitude of human afflictions previously mentioned are alleviated by dietary consumption of n3 LC-PUFAs. The diversity of these diseases implies that a general mode of action common to all cells is a strong possibility. Wassall *et al.* hypothesized that incorporation of DHA into the phospholipids of the plasma membrane and their lateral sequestration into domains from which cholesterol is excluded is the molecular origin of the responsible mechanism [4].

The role that membrane phospholipids fatty acids composition plays in signal transduction and receptor activity has been extensively pursued during the past 50 years [20]. Many studies have demonstrated that LC-PUFAs such as

AA, EPA and DHA are important structural components of membrane phospholipids and sphingolipids [19]. DHA has been shown to be rapidly incorporated into a variety of cells, primarily into phospholipids such as phosphatidylethanolamine, phosphatidylcholine and phosphatidylserine of the plasma membrane. The addition of acyl chain double bonds is generally assumed to increase fluidity and it may be surmised that a DHA-rich membrane should be exceptionally fluid. Many dietary studies have reported increases in membrane fluidity from animals fed DHA-rich fish oil diets, whereas a DHA-deficient diet resulted in a brush border membrane with decreased fluidity. Therefore, DHA-rich regions in membranes are thin with loose lipid packing. DHA 1) increases the area/head group, 2) increases permeability to water and other solutes, 3) increases acyl chain flexibility and dynamics, 4) disorders the bilayer in the region near the head group because double bonds extend closer to the glycerol backbone, and 5) increases packing free volume and induces negative curvature strain to proteins [3]. Because a fluctuating rough surface is incompatible with proximity to the rigid steroid moiety of cholesterol, DHA-containing phospholipids have poor affinity for the sterol domains within membranes. This fatty acid interacts poorly with cholesterol and tends to laterally separate into DHA-rich/sterol-poor regions [4] and may enhance formation of lateral domains [3]. They are also more "dynamic" than membrane regions composed of other fatty acids [4].

According to this more flexible structure that rapidly interconverts through a wide range of conformations, it distinguishes LC-PUFAs—most markedly DHA—from less unsaturated fatty acids [4]. Many reports have demonstrated that DHA-containing bilayers behave differently from bilayers containing the other major membrane PUFAs, e.g., arachidonic acid, further supporting the uniqueness that DHA is a main fatty acid from the nervous system. A simple elimination of one double bond from DHA producing DPA (22:5, n3) results in the loss of several behavioral features in animals [21].

Thus, the n3 and n6 fatty composition of biological membranes significantly influences physical properties of biological membranes such as the activity of signaling proteins, thereby altering protein function and trafficking, and vesicle budding and fusion [2,4]. However, more studies are necessary to fully elucidate the molecular mechanisms that n3 LC-PUFAs perform to regulate the cellular environment in the membrane microdomains.

Regulation of Gene Transcription

In addition to being vital components of membrane phospholipids and functioning in key steps of cell signaling, LC-PUFAs govern the expression of a wide array of genes through different transcription factors such as peroxisome proliferator-activated receptors (PPARs), sterol regulatory-element binding protein (SREBP), carbohydrate response-element binding protein (ChREBP) and nuclear factor κB (NFκB) mainly to regulate the transcription of a wide array of target genes including those encoding proteins involved with lipid and carbohydrate metabolism, thermogenesis, and cell differentiation (Fig. 2) [22,23,24].

a) PPARs

PPARs are well-described ligand-activated nuclear transcription factors that play important roles in cellular differentiation, cancer, insulin sensitization, atherosclerosis and several metabolic diseases [2]. Of the multiple mechanisms by which PUFAs regulate gene expression, fatty acid control of PPARs represents the one best understood. Because these transcription factors were activated by agents associated with peroxisomal proliferation, they were given the name PPARs [25]. PPARs are activated by LC-PUFAs. They are then able to form heterodimers with the retinoid X receptor (RXR). These dimers bind to PPAR responsive elements (PPRE) consisting of direct repeats of AGGTCA separated by one nucleotide (DR1) in the regulatory region of target genes and thus influence their expression. The predicted amino acids sequence for PPARα indicates that this receptor possesses structural characteristics of steroid receptors, i.e., a ligand-binding domain and a zinc finger DNA-binding domain [20].

There are three isoforms of PPAR: PPARα (NR1C1), PPARδ/β (NR1C2) and PPARγ (NR1C3) 1 and 2 that differ in ligand specificity, tissue distribution, and developmental expression. Expression of PPARα is greatest in tissues with active metabolism such as brown adipose tissue, liver, striated muscle and kidney. PPARδ has a very broad expression pattern that makes identification of its role more difficult. PPARγ is highly expressed in fat, colon, placenta and macrophages [26].

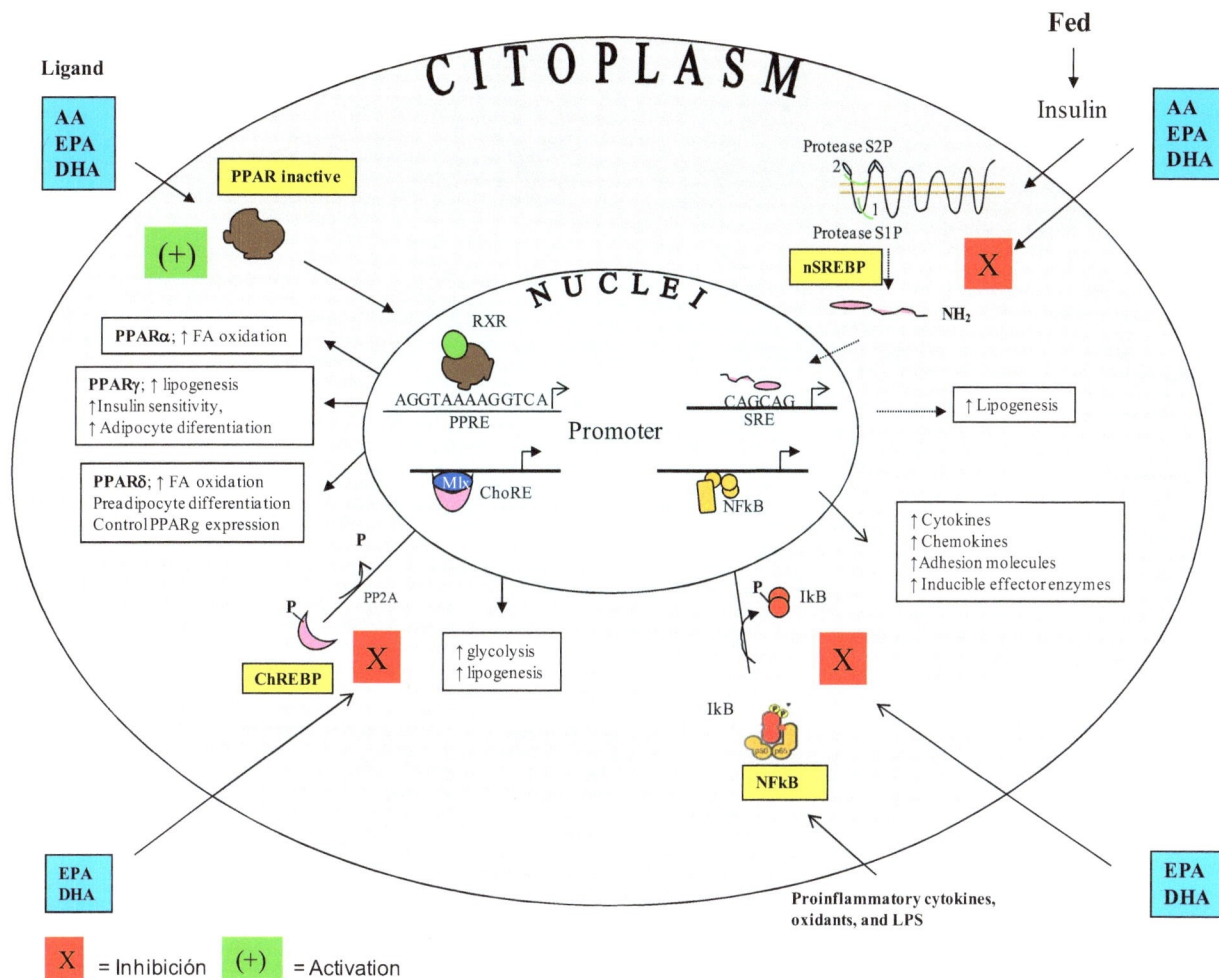

Figure 2: LC-PUFAs as regulators of transcription factors SREBP, PPAR, ChREBP and NFκB. X = Inhibiton (+) = Activation

Various PUFAs, especially those classified as n3 fatty acids, are natural ligands for activating PPARs [2]. Concerning activation potency, n3 fatty acids EPA and DHA are more potent as *in vivo* activators of PPARα than n6 fatty acids and can bind with sufficient affinities [2,27]. Certain fatty acids, however, are better than others at activating PPAR. Structural studies have established that EPA is both a ligand and a robust activator of PPAR. DHA and EPA, however, are weak PPARα activators because of their poor binding to the PPAR gene [28]. In general, all isoforms of PPAR are more responsive to 18–22 carbon n6 and n3 PUFAs than to saturated and monounsaturated fatty acids. The finding that polyenoic fatty acids are more potent activators of PPARs than are saturated and monounsaturated fatty acids parallels the metabolic findings that n6 and n3 PUFAs are more potent inducers of fatty acids oxidation and more potent suppressers of fatty acids and triacylglycerol synthesis [20].

This correlation has led to the proposal that PPARs coordinately regulate the expression of genes involved in the oxidation and synthesis of lipids. Numerous reports have established that the flanking regions of genes encoding carnitine *O*-palmitoyltransferase, acyl-CoA oxidase, mitochondrial hydroxymethylglutaryl-CoA synthase, fatty acyl-CoA synthetase, and mitochondrial uncoupling proteins contain PPRE [20]. Thus, a diet rich in 20- and 22-carbon PUFAs (i.e., PPAR ligand activators) significantly increases the expression of the aforementioned genes and that induction of these genes is associated with higher rates of fat oxidation and reduced body fat deposition [25].

Like all nuclear receptors, binding of the ligand (in this case the fatty acid) stimulates an exchange of co-activators for co-repressors on the chromatin-bound receptor and the recruitment of additional proteins involved in gene transcription such as RNA polymerase II. Numerous studies indicate that lipid activation of PPARs results in a dominant role in controlling genes involved in fatty acids oxidation, fatty acids desaturation and elongation, transport, and binding [28]. Therefore, it has been proposed that genes encoding the oxidative enzymes appear to be regulated by a common transcriptional factor. Also, because PPARs are lipid-activated transcription factors, they have often been proposed as the "master switches" that regulate the expression of enzymes involved in lipid synthesis and degradation [25].

PPARα

This isoform is abundantly expressed in tissues with a high capacity for fatty acids oxidation and activates a program of lipid-induced genes encoding proteins involved in FA uptake, activation and oxidation [29]. PPARα controls the expression of several genes involved in fatty acids metabolism from transport across the cell membrane, intracellular binding [liver fatty acids binding protein (FABP)], and formation of acyl-CoA to mitochondrial and peroxisomal β-oxidation as well as microsomal ω-oxidation. During fasting, when increased utilization of free fatty acids occurs, PPARα expression and activity promotes increased β-oxidation. Although PPARα mice are relatively healthy when fed *ad libitum*, they have very poor tolerance for fasting and develop hypoglycemia, hypoketonemia and hypothermia. In skeletal muscle, loss of PPARα is relatively mild, suggesting that PPARδ and perhaps other factors may compensate. However, PPARα is induced after exercise in humans as are some of its target genes [26].

PPARγ

Two isoforms of PPARγ (PPARγ1 expressed in colon, retina, spleen, hematopoietic cells, liver and skeletal muscle) and PPARγ2 (expressed in adipose tissue) are able respond to the same signals and activate the same target genes: aP2, LPL, ACS and CD36 [26]. PPARγ controls fat metabolism by regulating genes involved in lipogenesis, insulin sensitivity, and adipocyte differentiation. These effects underlie the use of thiazolidinediones (pharmacological ligand), which bind and activate PPARγ to treat insulin-resistant type II diabetes [30,31]. PPARγ is adipose enriched and promotes lipid synthesis and storage in mature adipocytes. PPARγ is both necessary and sufficient for preadipocyte differentiation to mature adipocytes (and thus the capacity for triglycerides storage). PPARγ regulates at least two programs of gene expression during adipogenesis in 3T3-L1 preadipocytes. One consists of classic adipogenic genes (FABP4/aP2, adiponectin, and perilipin). Following its induction, this program continues to be expressed throughout terminal adipogenesis and in mature adipocytes. The other program includes a diverse array of genes, one of which is fibroblast growth factor 21 (FGF21), a newly identified molecule considered to have therapeutic potential for diabetes. FGF21 expression in adipocytes is selectively activated by synthetic PPARγ ligands. Although the role of PPARγ is not fully understood, it has been proposed that it is also involved in the regulation of Tg storage in mature adipocytes [29].

PPARδ

This subtype, also known as PPARβ, has the broadest expression of the PPAR isotypes, implying a more general cell function that is not fully understood, although its biological function has recently been investigated through genetic manipulation [32]. PPARδ plays a key role in lipid metabolism of extrahepatic tissues, muscle fiber-type remodeling, lipoprotein metabolism and glucose uptake [33] and antagonizes multiple proinflammatory pathways [32]. Although it is expressed in adipose tissue, it is not directly involved in preadipocyte differentiation but is implicated in the control of preadipocyte proliferation and PPARγ gene expression [29]. The actions of PPARδ appear to be more similar to PPARα than PPARγ in favoring the oxidation of fats. In cell culture, UCP2, UCP3, H-FABP, FAT/CD36, LPL, ACS and CPT1 were demonstrated to be PPARδ target genes. Restricted activation of this receptor in either muscle or adipose tissue results in a lean phenotype due to increased fatty acids β-oxidation. This hypothesis has been demonstrated because long-term treatment with a PPARδ ligand GW501516 causes dramatic weight loss accompanied by improvements in lipoprotein profiles [32]. Also, overexpression of PPARδ in fat produced a lean phenotype due to increased oxidation of fats in adipocytes [26].

b) SREBPs

SREBPs are a family of transcription factors first isolated as a result of their properties for binding to the sterol regulatory element (SRE). There are three members of the SREBP family: 1a, 1c, and 2. SREBP-1a and -1c are derived from the same gene, but different to SREBP-1 c the amino-terminus of the SREBP-1a protein has a 24-amino acids residue and SREBP-1a and -1c use different promoter sites [34]. SREBP-1c is expressed in most mouse and human tissues with high levels in macrophages, liver, white adipose tissue, adrenal gland and brain and constitutes 90% of the SREBP-1 found *in vivo*. SREBP-1c controls hepatic and total body cholesterol and fatty acids synthesis and also functions to regulate multiple facets of fatty acids synthesis and VLDL assembly. SREBP-2 plays a major role in the regulation of cholesterol synthesis and uptake [28]. SREBPs are synthesized as 125-kDa precursor proteins that contain two transmembrane domains for insertion into the endoplasmic reticulum (ER) membrane [34]. Precursor SREBPs are escorted from the ER to the Golgi complex by SREBP cleavage-activating protein (SCAP) for proteolytic processing. Proteases in the Golgi, i.e., site 1 and site 2 protease, cleave the SREBP precursor to generate the mature nuclear form of SREBP (nSREBP). The N-terminal domain, a 68-kDa helix–loop–helix leucine zipper transcription factor (nSREBP), is transported to the nucleus as a dimer via importin β. In nuclei, SREBPs bind to SRE as dimers in promoters of target genes. Once bound, SREBPs recruit co-activators to the promoter and stimulate gene transcription [28].

Sterols regulate nSREBP levels by controlling the proteolytic processing step at the level of the ER and Golgi, not the proteasomal degradation of the nuclear form of SREBP. Sterols induce the ER resident proteins, Insig-1 or Insig-2, to bind SCAP. The Insig–SCAP–SREBP complex is retained in the ER, preventing its movement to the Golgi for cleavage and maturation to nSREBP. In this manner, cholesterol functions as a feedback inhibitor of cholesterol synthesis by preventing SREBP-2 from accumulating in nuclei and inducing expression of key genes involved in cholesterol synthesis. Whereas SREBP-1 and SREBP-2 are structurally similar, their regulation in the liver by nutrients and hormones is quite different because the nuclear abundance is controlled by several mechanisms that include dietary PUFAs [28].

Polyunsaturated fatty acids n3 and n6 are key regulators of SREBP-1c activation and consequently of fatty acids synthesis in liver. PUFAs are well-established suppressors of SREBP-1 but not SREBP-2 nuclear abundance. Enhanced levels of endogenously synthesized or externally provided PUFAs in liver repress SREBP-1c expression and activity, which result in a downregulation of both *de novo* lipogenesis and PUFAs synthesis through this negative feedback system [10,35]. Dietary PUFAs inhibit hepatic lipogenesis by suppressing the transcription of hepatic genes encoding glycolytic and lipogenic enzymes (fatty acid synthase (FAS)), acetyl-CoA carboxylase (ACC), stearoyl-CoA desaturase (SCD), malic enzyme, L-pyruvate kinase, glucose-6-phosphate dehydrogenase, citrate lyase, stearoyl CoA desaturase, glucokinase) [25], ABCG1, which mediates cholesterol efflux [2], and desaturases and elongases, are enzymes participating in LC-PUFAs synthesis [24,28]. The effects of various fatty acids on nSREBP-1 have been examined. PUFAs inhibit SREBP-1 gene transcription and induce mRNA SREBP-1 instability that, in turn, leads to a reduction in the amount of membrane-anchored precursor SREBP-1 protein and inhibits proteolytic release of mature SREBP from its membrane-anchored precursor [36]. It has been demonstrated that fish oil (rich in DHA) and diets rich in LA n6 or EPA n3 and DHA n3 reduced hepatic premature and nuclear SREBP-1 up to 85% and 60%, respectively [2]. Recently, it was reported that DHA was the most potent suppressor of SREBP-1 nuclear abundance compared with AA n6 and EPA n3; thus, DHA n3 is likely the major regulator of nSREBP-1 abundance *in vivo* [37].

c) ChREBP

Recently, ChREBP was shown to play a pivotal role in the induction of glycolytic and lipogenic genes by glucose due to its capacity to bind to the carbohydrate-responsive element (ChoRE) present in promoters of these target genes [22]. Binding of ChREBP to these regulatory elements, however, requires a Max-like protein X (Mlx). Neither insulin nor glucose regulates nuclear Mlx content; nuclear Mlx content appears constitutive. Glucose stimulates the accumulation of ChREBP, a newly recognized transcription factor expressed in liver, and is responsive to the nutritional state [38]. ChREBP is a large transcription factor of the Mondo family consisting of 865 amino acids with a weight of 96,500 kDa. It contains several functional domains including nuclear export signal site, nuclear import signal sites, a DNA binding bHLH/Zip domain, and proline-rich regions implicated in protein–protein interactions including multiple phosphorylation sites. Previous studies have shown that the N-terminal region of ChREBP (residues 1–251) interacts directly with a dimeric form of the protein and is responsible for subcellular localization, whereas the C-terminal region of ChREBP forms a complex with Mlx to bind to various DNA sites, thereby activating the transcriptional activity [39]. ChREBP binds ChoRE and E-boxes in promoters of responsive genes like SCD1, l-PK, FAS, and ACC. However, Zhao *et al.* were unable to show a significant stimulatory effect of the transcription factor on all the ACC gene promoters, except that the ChREBP/Mlx heterodimer tended to enhance ACC1 PI (promoter I). One possibility is that ChREBP/Mlx alone is not sufficient, and the co-existence of other factors such as SREBP-1c and LXR is needed to cause a significant effect [38]. Additional studies are needed to clarify the role of ChREBP in ACC gene regulation. Regarding I-PK, whereas increased glucose metabolism, possibly through the pentose phosphate pathway, is important for the induction of hepatic I-PK, PUFAs inhibit its transcription. Functional mapping analyses have shown that I-PK promoter contains PUFAs response element and is colocalized with the ChoRE binding site. However, despite several years of investigation, the identity of the transcription factor(s) involved in mediating the negative effects of PUFAs on the I-PK gene remains unclear [22].

Stimulation of ChREBP by glucose occurs at two levels. High glucose and insulin concentrations stimulate ChREBP gene expression and also stimulate its translocation from the cytosol to the nucleus, thereby increasing DNA-binding/transcriptional activity of ChREBP [22]. With an excess of glucose, hepatic glucose metabolism through the pentose phosphate pathway increases, leading to elevated levels of xylulose-5-phosphate (X5P). X5P is an activator of protein phosphatases 2A (PP2A). PP2A dephosphorylates ChREBP, allowing ChREBP to accumulate in nuclei. The influx of ChREBP into the nuclei likely triggers binding of ChREBP to its obligate heterodimer partner, MLX, resulting in enhanced occupancy of ChoRE with nascent ChREBP/MLX. This heterodimer binds ChoRE in promoters of target genes [28].

The fact that the DNA-binding activity of ChREBP in nuclear extract of livers from rats fed a high-fat diet is decreased compared with that in rats fed a high-carbohydrate diet suggests that ChREBP may be intimately involved in fatty acids inhibition of glycolysis and lipogenesis. Dentin *et al.* hypothesized that the suppressive effect of PUFAs on glycolytic and lipogenic gene transcription may be partially mediated through a decrease in ChREBP gene expression and/or a defect in its nuclear translocation [22]. This negative effect is not likely fatty acids specific because it was reported that there was no differential effect in potency of 20:4n-6 (AA), 20:5n-3 (EPA) or 22:6n-3 (DHA) on the control of glycolytic gene transcription, ChREBP or MLX nuclear abundance [28]. Therefore, unlike PPARα and SREBP-1 (see above), PUFAs regulation of ChREBP and MLX nuclear abundance is less responsive to the type of PUFA. The outcome of these studies provided novel evidence of how PUFAs controls glycolysis, in particular, I-PK. However, it is necessary to continue exploring the molecular basis for LC-PUFAs control of MLX or ChREBP nuclear abundance.

d) NFκB

NFκB is a transcriptional activator protein and plays an important role in controlling inflammatory gene activation. NFκBs are comprised of dimers normally confined to the cytoplasm through their association with the inhibitory subunit IκBs, which mask their nuclear localization sequence. When cells are activated by proinflammatory cytokines, oxidants, and LPS, IκBs are rapidly phosphorylated and degraded to free NFκBs, which migrate to the nucleus where they bind to cognate DNA binding sites and activate gene transcription [40]. Activation of NFκB results in the induction of a large number of genes involved in the regulation of a wide variety of biological responses including anti-apoptotic genes, cell cycle regulatory genes, genes encoding adhesion molecules, chemokines, inflammatory cytokines, genes involved in metastases, cyclooxygenase, and VEGF. Therefore, NFκB controls several cytokines (e.g., IL-1, IL-2, IL-6, IL-12, TNF-α), chemokines (e.g., IL-8, MIP-1a, MCP1), adhesion molecules (e.g., ICAM, VCAM and E-selectin), and inducible effector enzymes (e.g., iNOS and COX-2). Regulation of the transcription factor NFκB activity occurs at several levels including controlled cytoplasmic-nuclear shuttling and modulation of its transcriptional activity [41]. n3 LC-PUFAs directly inhibit NFκB activity. EPA significantly decreases IκB phosphorylation at serine 32 and consequently prevents NFκB migration to the nucleus to bind to the specific consensus sequence from cytokines (e.g., TNFα) in cultured pancreatic cells and human monocytes. This demonstrated a mechanism for proinflammatory cytokine transcription inhibition, in part through inactivation of the NFκB signal transduction pathway secondary to inhibition of IκB phosphorylation [23,42]. This is supported by the finding of Hudert *et al.* where transgenic mice that endogenously biosynthesize n3 PUFAs are protected from colitis through a decrease in NFκB activity. The authors demonstrate that an increased n3 LC-PUFAs status generates higher levels of bioactive n3 LC-PUFAs-derived lipid mediators (resolvins and protectins), which may, on one hand, suppress the inflammatory response and, on the other hand, enhance mucoprotection (defense of intestinal mucosa) and is thereby protected against inflammation and injury. A recent study also suggests that DHA has greater anti-inflammatory effects on the protein levels of the components of the NFκB transcriptional system than EPA [2].

In summary, there are harmonic molecular mechanisms where LC-PUFAs regulate cell functions mainly by modulation of signal transduction via effect of bioactivity on cell membranes and regulation of gene transcription to maintain cell homeostasis. Experimental evidence indicates that n3 fatty acids are more potent cellular regulators than n6 fatty acids. One good starting point is that it is well known that LC-PUFAs control some key cellular molecular mechanisms such as lipid and carbohydrate metabolism and inflammation through master transcription factors regulators. However, more studies are required to elucidate these actions and to gain a better understanding about the beneficial role of LC-PUFAs in the pathogenesis of various diseases. Therefore, integrative analysis including nutritional, biochemical, genetic and immunological studies may provide information about the identification of specific molecular mechanisms involved in the beneficial effects of n3 LC-PUFAs such as DHA and EPA intake and their metabolic derivates on health promotion or disease burden.

REFERENCES

[1]   El-Badry, A. M.; Graf, R.; Clavien, P. A. Omega 3–Omega 6: what is right for the liver? *J. Hepatol.,* **2007**, *47*, 718–725.

[2]   Schmitz, G.; Ecker, J. The opposing effects of n-3 and n-6 fatty acids. *Prog. Lip. Res.,* **2008**, *47*, 147–155.

[3]   Stillwell, W.; Wassall, S. R. Docosahexaenoic acid: membrane properties of a unique fatty acid. *Chem. Phys. Lip,.* **2003**, *126*, 1–27.

[4]   Wassall, S. R.; Stillwell, W. Docosahexaenoic acid domains: the ultimate non-raft membrane domain. *Chem. Phys. Lip.,* **2008**, *153*, 57–63.

[5]   Anderson, B. M.; Ma, D. W. L., Are all n-3 polyunsaturated fatty acids created equal?. *Lipids Health Dis.,* **2009**, *8*, 33-53.

[6]   Sealls, W.; Gonzalez, M.; Brosnan, M. J.; Black, P. N.; DiRusso, C. C. Dietary polyunsaturated fatty acids (C18:2 ω6 and C18:3 ω3) do not suppress hepatic lipogenesis. *Biochim. Biophys. Acta.,* **2008**, *1781*, 406–414.

[7]   Dreesen, T. D.; Adamson, A. W.; Tekle, M.; Tang, Ch.; Cho, H. P.; Clarke, S. D.; Gettysa, T. W. A newly discovered member of the fatty acid desaturase gene family: a non-coding, antisense RNA gene to D5-desaturase. *Prostag. Leukot. Essent. Fatty. Acids.,* **2006**, *75*, 97–106.

[8]   Leonard, A. E.; Pereira, S. L.; Sprecher, H.; Huang, Y. S. Elongation of long-chain fatty acids. *Prog. Lip. Res.,* **2004**, *43*, 36-54.

[9]   Leonard, A. E.; Kelder, B.; Bobik, E. G.; Chuang, L. T.; Lewis, C. J.; Kopchick, J. J.; Mukerji, P.; Huang, Y. S. Identification and expression of mammalian long-chain PUFA elongation enzymes. *Lipid,* **2002**, *37*, 733-740.

[10]   Qin, Y.; Dalen, K. T.; Gustafsson, Jan-Ake; Nebb, H. Regulation of hepatic fatty acid elongase 5 by LXRα–SREBP-1c. *Biochim. Biophys. Acta,* **2009**, *1791*, 140–147.

[11]   Cho, H. P.; Nakamura, M. T.; Clarke, S. D. Cloning, expression, and fatty acid regulation of the human D-5 desaturase. *J. Lipid Res.,* **1999**, *274*, 37335–37339.

[12]   Shanklin, J.; Whittle, E.; Fox, B. G. Eight histidine residues are catalytically essential in a membrane-associated iron enzyme, stearoyl-CoA desaturase, and are conserved in alkane hydroxylase and xylene monooxygenase. *Biochemistry,* **1994**, *33*, 12787-12794.

[13]   Hashimoto, K.; Yoshizawa, A. C.; Okuda, S.; Kuma, K.; Goto, S.; Kanehisa, M. The repertoire of desaturases and elongases reveals fatty acid variations in 56 eukaryotic genomes. *J. Lipid Res.,* **2008**, *49*, 183-91.

[14]   Jakobsson, A.; Westerberg, R., Jacobsson, A. Fatty acid elongases in mammals: their regulation and roles in metabolism. *Prog. Lip. Res.,* **2006**, *45*, 237–249.

[15]   Moon, Y. A.; Hammer, R. E.; Horton, J. D. Deletion of ELOVL5 leads to fatty liver through activation of SREBP-1c in mice, *J. Lipid Res.,* **2009**, *50*, 412-23.

[16]   Wang, Y.; Botolin, D.; Christian, B.; Busik, J.; Xu, J.; Jump, D. B. Tissue-specific, nutritional, and developmental regulation of rat fatty acid elongases. *J. Lipid Res.,* **2005**, *46*, 706–715.

[17]   Nakamura, M. T.; Cho, H. P.; Xu, J.; Tang, Z.; Clarke, S. D. Metabolism and functions of highly unsaturated fatty acids: an update. *Lipids,* **2001**, *36*, 961–964.

[18]   Salem, N.; Litman, B.; Kim, H. Y.; Gawrisch, K., Mechanisms of action of docosahexaenoic acid in the nervous system. *Lipids,* **2001,** *36*, 945–959.

[19]   Sampath, H.; Ntambi, J. M. Polyunsaturated fatty acid regulation of genes of lipid metabolism. *Annu. Rev. Nutr.,* **2005**, *25*, 317-340.

[20]   Clarke, S. D.; Thuillier, P.; Baillie, R. A., Sha, X. Peroxisome proliferator-activated receptors: a family of lipid-activated transcription factors. *Am. J. Clin. Nutr.,* **1999**, 70, 566–571.

[21]   Eldho, N. V.; Feller, S. E.; Tristram-Nagle, S.; Polozov IV.; Gawrisch, K. Polyunsaturated docosahexaenoic vs. docosapentaenoic acid—differences in lipid matrix properties from the loss of one double bond. *J. Am. Chem. Soc.*, **2003**, *125*, 6409–6421.

[22]   Dentin, R.; Benhamed, F.; Pégorier, J. P.; Foufelle, F.; Viollet, B.; Vaulont, S.; Girard, J.; Postic, C. Polyunsaturated fatty acids suppress glycolytic and lipogenic genes through the inhibition of ChREBP nuclear protein translocation. *J. Clin. Invest.*, **2005**, *115*, 2843–2854.

[23]   Novak, T. E.; Babcock, T. A.; Jho, D. H.; Helto, W. S.; Espat, N. J. NF-κB inhibition by ω-3 fatty acids modulates LPS-stimulated macrophage TNF-α transcription. *Am. J. Physiol. Lung Cell Mol. Physiol.*, **2003**, *284*, L84–L89.

[24]   Nara, T. Y.; He, W. S.; Tang, Ch.; Clarke, S. D.; Nakamura, M. T., The E-box-like sterol regulatory element mediates the suppression of human D-6 desaturase gene by highly unsaturated fatty acids. *Biochim. Biophy. Res. Commun.*, **2002**, *296*, 111–117.

[25]   Xu, J.; Nakamura, M. T.; Cho, H. P.; Clarke, S. D. Sterol regulatory element binding protein-1 expression is suppressed by dietary polyunsaturated fatty acids. *J. Biol. Chem.*, **1999**, *274*, 23577–23583.

[26]   Alaynick, W. A. Nuclear receptors, mitochondria and lipid metabolism. *Mitochondrion*, **2008**, *8*, 329–337.

[27]   Gani, O. A.; Sylte, I., Molecular recognition of docosahexaenoic acid by peroxisome proliferator-activated receptors and retinoid-X receptor α. *J. Mol. Graph. Model.*, **2008**, *27*, 217–224.

[28]   Jump, D. B.; Botolin, D.; Wang, Y.; Xua, J.; Demeure, O.; Christian, B., Docosahexaenoic acid (DHA) and hepatic gene transcription. *Chem. Phys. Lipids*, **2008**, *153*, 3–13.

[29]   Sugden, M. C.; Zariwala, M. G.; Holness, M. J. PPARs and the orchestration of metabolic fuel selection. *Pharmacol. Res.*, **2009**, *60*, 141–150.

[30]   Feige, J. N.; Gelman, L.; Michalik, L.; Desvergne, B.; Wahli, W. From molecular action to physiological outputs: peroxisome proliferator-activated receptors are nuclear receptors at the crossroads of key cellular functions. *Prog. Lipid Res.*, **2006**, *45*, 120–159.

[31]   Michalik, L.; Auwerx, J.; Berger, J. P.; Chatterjee, V. K.; Glass, C. K.; Gonzalez, F. J.; Grimaldi, P. A.; Kadowaki, T.; Lazar, M. A.; O'Rahilly, S.; Palmer, C. N.; Plutzky, J.; Reddy, J. K.; Spiegelman, B. M.; Staels, B.; Wahli, W. International union of pharmacology. LXI. Peroxisome proliferator activated receptors. *Pharmacol. Rev.*, **2006**, *58*, 726–741.

[32]   Barish, G. D.; Atkins, A. R.; Downes, M.; Olson, P.; Chong, L. W.; Nelson, M.; Zou, Y.; Hwang, H.; Kang, H.; Curtiss, L.; Evans, R. M.; Lee, Ch. PPARδ regulates multiple proinflammatory pathways to suppress atherosclerosis. *PNAS*, **2008**, *105*, 4271–4276.

[33]   Zeng, L.; Geng, Y.; Tretiakova, M.; Yu, X.; Sicinski, P.; Kroll, T. G. PPARδ induces cell proliferation by a cyclin E1-dependent mechanism and is upregulated in thyroid tumors. *Cancer Res.*, **2008**, *68*, 6578–6586.

[34]   Xu, J.; Teran-Garcia, M.; Park, J. H. Y.; Nakamura, M. T.; Clarke, S. D. Polyunsaturated fatty acids suppress hepatic *Sterol Regulatory Element-binding Protein-1* expression by accelerating transcript decay. *J. Biol. Chem.*, **2001**, *276*, 9800–9807.

[35]   Tam, P. S. Y.; Sawada, R.; Cui, Y.; Matsumoto, A.; Fujiwara, Y. The metabolism and distribution of docosapentaenoic acid (n-6) in the liver and testis of growing rats. *Biosci. Biotechnol. Biochem.*, **2008**, 72, 2548-2554.

[36]   Tang, Ch.; Cho, H. P.; Nakamura, M. T.; Clarke, S. D. Regulation of human Δ-6 desaturase gene transcription: identification of a functional direct repeat-1 element. *J. Lipid Res.*, **2003**, *44*, 686–695.

[37]   Botolin, D.; Wang, Y.; Christian, B.; Jump, D. B. Docosahexaneoic acid [22:6, n-3] stimulates rat hepatic sterol regulatory element binding protein-1c (SREBP-1c) degradation by an Erk- and 26S proteasome-dependent pathway. *J. Lipid Res.*, **2006**, *47*, 181–192.

[38]   Zhao, L. F.; Iwasaki, Y.; Zhe, W.; Nishiyama, M.; Taguchi, T.; Tsugita, M.; Machiko, Kambayashi.; Hashimoto, K.; Terada, Y. Hormonal regulation of acetyl-CoA carboxylase isoenzyme gene transcription. *Endocr. J. Adv. Public.*, **2010**, 57, 317-324.

[39]   Fukasawa, M.; Ge, Q.; Wynn, R. M.; Ishii, S.; Uyeda, K. Coordinate regulation/localization of the carbohydrate responsive binding protein (ChREBP) by two nuclear export signal sites: discovery of a new leucine-rich nuclear export signal site. *Biochem. Biophys. Res. Commun.*, **2010**, *391*, 1166–1169.

[40]   Lo, Ch. J.; Chiu, K. C.; Fu, M.; Lo, R.; Helton, S. Fish oil decreases macrophage tumor necrosis factor gene transcription by altering the NFκB activity. *J. Surg. Res.*, **1999**, *82*, 216–221.

[41]   Ghosh, S.; Karin, M. Missing pieces in the NF-kappaB puzzle. *Cell*, **2002**, *109*, S81–S96.

[42]   Zhao, Y.; Joshi-Barve, S.; Barve, S.; Chen, L. H. Eicosapentaenoic acid prevents LPS-induced TNF-alpha expression by preventing NFkappaB activation. *J. Am. Coll. Nutr.*, **2004**, *23*, 71–78.

CHAPTER 2

Role of Long-Chain Polyunsaturated Fatty Acids in Pregnancy and Lactation: Fetal and Infant Growth

Maricela Rodríguez-Cruz M. Sc., Ph. D.

Laboratorio de Biología Molecular, Unidad de Investigación Médica en Nutrición, Hospital de Pediatría, Centro Medico Nacional Siglo XXI, IMSS, México City, México

Abstract: The purpose of this review is to provide an overview regarding the role of ω-3 and ω-6 long-chain polyunsaturated fatty acids (LC-PUFAs), arachidonic acid (AA) and docosahexaenoic acid (DHA) on normal growth and maturation of the central nervous system and retina of the fetus, newborn and infant. Numerous studies have shown that DHA is associated with higher scores on tests of visual and neural development in infants and children. We also present progress concerning the molecular mechanism triggered during pregnancy and lactation to support LC-PUFAs requirements. During pregnancy, the fetus demands LC-PUFAs, which are provided through placental transfer. Placental transfer of fatty acids involves a multi-step process of uptake and translocation facilitated by specific proteins that favor DHA and AA over other fatty acids. After birth, the newborn acquires the LC-PUFAs from milk or formula. LC-PUFAs from cord blood and breast milk are acquired from the maternal diet, mobilized from reserves, or synthesized *de novo* in the maternal organism from the precursors linoleic acid (LA) and linolenic acid (LNA). The mother adapts her metabolism to support this draining of LC-PUFAs through mammary tissue, using a high rate of dietary uptake and allowing the expression of enzymes responsible for LC-PUFAs synthesis. We have demonstrated that mammary tissue, together with the liver, plays an important role in the synthesis of ω-3 and ω-6 LC-PUFAs to supply to the product in pregnancy and lactation.

INTRODUCTION

Arachidonic acid [AA; 20:4(ω-6)] and docosahexaenoic acid [DHA; 22:6(ω-3)] are the most abundant long-chain polyunsaturated fatty acids (LC-PUFAs) in the membrane phospholipids of neural tissues, including brain. These fatty acids have been of particular interest in nutrition during pregnancy because of their potential effects on pregnancy outcomes, maternal health and development of the offspring [1]. Especially during rapid neurodevelopment of the fetus in the last trimester of pregnancy, in the early postnatal period and maybe during infancy (at least the first 4 years of life) AA and DHA accumulate in large amounts in neural tissues, where they serve as important structural and functional components in the development of neural and synaptic networks [2]. This challenge is met by increased extraction of LC-PUFAs via the placenta and during lactation, women continue the transfer of their own LC-PUFAs to their infants, but this does not compromise the restoration of their relative plasma phospholipid AA concentrations [2]. LC-PUFAs from maternal plasma phospholipids or breast milk vary considerably, mainly as a consequence of differences in maternal intake of these fatty acids [3]. The influence of maternal diet on fatty acids milk composition cannot be adequately addressed without considering maternal nutritional status prior to and during pregnancy. This pronouncement has been well demonstrated in different studies, e.g., fish-eating populations have higher breast milk DHA concentrations than do populations who do not consume marine foods [4]. In fact, higher intakes of DHA increase maternal/fetal transfer of DHA in breast milk during lactation. Dietary DHA is also readily taken up and incorporated into the developing brain and retina [5]. The importance of LC-PUFAs for neural and visual development was demonstrated by numerous studies that evaluated the effect of DHA status on the developing visual system. These studies have shown that the DHA status of infants at birth was related to the maturity of visual-evoked potential and improved eye/hand coordination [6]. Other studies have shown that children born to mothers supplemented with cod liver oil (rich in EPA and DHA) during pregnancy and lactation scored higher on the Mental Processing Composite of the Kaufmann Assessment Battery for Children at 4 years of age than did children born to women in the group using corn oil [7]. In contrast, studies in different countries have linked low intakes of fish (the major source of DHA) in pregnant women to low blood levels of DHA

*Address correspondence to Dr. Maricela Rodríguez-Cruz; Apartado postal C-029 C. S.P.I. "Coahuila" Coahuila No. 5, Col. Roma 06703. México, D. F., México. Tel. +52 56276900, ext. 22483, 22484. Fax +52 56276944; E-mail: maricela.rodriguez.cruz@gmail.com

during pregnancy or in infants at birth. Infants exposed to breast milk with low DHA had lower test scores on mental aptitude and less motor and visual system development, with these effects extending into later childhood [5]. Therefore, the fetus, neonate and infant should receive ω-3 and ω-6 LC-PUFAs through placenta, breast milk or formula.

This review represents an overview regarding the role of ω-3 and ω-6 LC-PUFAs, arachidonic acids (AA) and docosahexaenoic acid (DHA) on normal growth and maturation of the central nervous system and retina of the fetus, newborn and infant. We review the progress about the molecular mechanisms triggered during pregnancy and lactation to support LC-PUFAs requirements.

PERINATAL PHYSIOLOGICAL ADAPTATIONS OCCURRING IN THE MOTHER TO SUSTAIN FETAL GROWTH

Human reproduction and lactation are physiological functions vital for procreation and survival of the species. Pregnancy is a period of characteristic hormonal and metabolic changes that affect lipid and fatty acids metabolism [8,9]. During the first two trimesters of pregnancy there is an increase in maternal lipogenesis and in adipose tissue stores. However, during the third trimester, increasing serum concentrations of prolactin, cortisol, glucagon and chorionic somatomammotropin, which exert anti-insulinogenic and lipolytic effects, result in enhanced lipolysis in adipose tissue and, therefore, increased plasma levels of nonesterified fatty acids (NEFAs) [9]. NEFAs are used as energy sources by maternal peripheral tissues during fasting [9] and as a direct source of fatty acids to the placenta [10]. Therefore, pregnancy is generally associated with hyperlipidemia, which is thought to supply the maternal demand of lipids and essential fatty acids for energy and structural use by the growing fetus [11]. Thus, pregnancy and lactation are periods of high nutritional demands. Energy and nutrient requirements are elevated because of increased needs for the growing tissues of the placenta and fetus, as well as for milk production. These periods are characterized by increased nutritional risks for the mother and the growing fetus or infant [12].

During pregnancy, the maternal organism demands a substantial number of essential fatty acids, predominantly ω-3 and ω-6 LC-PUFAs because of their potential effects on maternal health and development of the offspring. These include normal oxidation for energy and physical requirements of the mother and accretion by the fetus. This proposition was supported from concentrations of plasma phospholipid-associated fatty acids that also rise during pregnancy, particularly DHA [1,11]. This is in agreement with postmortem studies of fetuses, stillbirths, and preterm infants. It was estimated that the fetus accretes ~50–60 mg/kg/day for ω-3 and 400 mg/kg/day for ω-6 fatty acids during the last trimester of pregnancy and during the early weeks of life, most of which is DHA [10,13,14]. This is equivalent to a dietary DHA content of ~1% of total fatty acids [15].

The demand for LC-PUFAs such as DHA ω-3 and AA increases during the last trimester of pregnancy and LC-PUFAs are transferred from the mother to the fetus [16,17]. These are obtained from the maternal diet and/or synthesized by the mother from the essential fatty acids, alpha linolenic acid (α–LNA, ω-3) and linoleic acid (LA, ω-6), respectively [10,18]. Although elongation and desaturation enzymes for LC-PUFAs synthesis are present in the fetal liver microsomes from as early as 17 weeks of gestation [19], activity appears to be low before birth. Thus, DHA and AA synthesis by the fetus is limited [20]. Therefore LC-PUFAs accumulated by the fetus *in utero* are derived predominantly through placental transfer, with the amounts in cord blood influenced by the maternal diet [6,19]. Placental and fetal demands may contribute to the decline of ω-6 and ω-3 LC-PUFAs proportions (g/100 g of total fatty acids) in maternal plasma phospholipids and erythrocyte membrane phospholipids during gestation [21].

Fetal metabolism and, consequently, fetal growth directly depend on the nutrients crossing the placenta. Therefore, the mother adapts her metabolism in order to support the continuous draining of substrates [22]. Some studies suggest adaptive metabolic mechanisms in pregnancy and lactation, such as increased synthetic capacity in liver and mammary gland [23], to metabolize LA and LNA to AA and DHA, respectively, and a preferential use of DHA stored in adipose tissue [24,25]. Also, it may be that increased dietary intake is necessary to maintain optimal function in women with habitually low dietary intakes [26]. Furthermore, pregnancy results in a state of amenorrhea, which is a form of nutrient conservation because there is no loss of LC-PUFAs via blood and other cellular material by menstruation [27].

Molecular Mechanism Involved in Supplying LC-PUFAs During Pregnancy and Lactation

Maternal LC-PUFAs are acquired from the diet, mobilized from adipose tissue reserves or synthesized *de novo* in the maternal organism from the essential PUFAs LA and LNA [28,29]. The mother provides LC-PUFAs to the product via different molecular mechanisms. We recently approached this hypothesis by identifying some of the metabolic adaptations that maternal organism uses to provide AA and DHA. Although the liver is considered the primary site of LC-PUFAs synthesis, we have proposed that lactating mammary tissue may participate in this process. The mammary gland is not considered a body pool for LA and AA; however, we have demonstrated that lactating mammary tissue has a high rate of dietary LA uptake and expresses desaturases, key enzymes responsible for AA and DHA synthesis. Δ6D and Δ5D expression was even higher in animals fed diets with a low LA content [23]. This expression was in accordance with the higher endogenous synthesis of AA [30]. Based on the fact that AA and DHA requirement increases during the last trimester of pregnancy, we evaluated elongases (Elovl 2 and 5) and desaturases (Δ6D and Δ5D) expression involved in the synthesis of AA and DHA during pregnancy and lactation. We identified the transcript of desaturases and elongases throughout pregnancy and lactation. These data suggest a significant role for mammary gland in the production of LC-PUFAs during pregnancy. Therefore, additional research is necessary to determine if this expression is related to different stages of pregnancy and/or lactation. With these data, we suggested that one of the molecular mechanisms involved in LC-PUFAs synthesis is that liver and mammary gland act in concert as a major producer of these fatty acids in a stage-specific manner for supporting pregnancy and lactation (Rodriguez-Cruz *et al.*, unpublished data). Thus, this tissue may temporarily store dietary LA before being secreted into milk or metabolized to AA and, therefore, represents an important source of these fatty acids for the newborn [30]. Also, it is important to consider that our tracer studies and studies by other investigators have indicated that the vast majority of milk essential fatty acids is not derived directly from the diet but from maternal stores [25,31]. We have proposed that <70% of LA and 90% of AA in milk are not derived directly from intestinal absorption. In addition, only a minor fraction of milk AA stems from its detectable conversion from dietary LA, which indicates that slowly turning-over maternal body pools are the major sources of milk LA and AA [25]. However, it is important to consider that AA may be subjected to a large distribution volume because of its high tendency to become incorporated into membrane phospholipids and, to a lesser extent, in plasma and adipose tissue triacylglycerols [32,33].

PLACENTAL FATTY ACIDS TRANSFER AND THE SECRETION OF FATTY ACIDS INTO HUMAN MILK

Placental Polyunsaturated Fatty Acids Transfer

Before birth, all of the ω-6 and ω-3 fatty acids accumulated by the fetus must originate from the maternal circulation through placental transfer. To establish the proportions fed to the fetus by the placenta, analyses of maternal and cord blood have been done at term and at midterm. Some studies using erythrocyte membrane [34] and plasma phospholipids [35] in the placenta and the umbilical cord blood have shown that AA and DHA contents are higher, whereas LA is lower in newborn infants than in their mothers. Full-term infant levels of LA, LNA and LC-PUFAs are associated with maternal levels [34,36]. Also, studies in midterm abortions may elucidate the events occurring close to the birth of a preterm baby. The transplacental gradient in proportions of AA and DHA appears to be greater at term (gestation weeks 30–38) than at midterm. This difference is consistent with the finding that, whereas DHA is mobilized at the beginning of pregnancy, its plasma concentrations fall toward the end, raising the suggestion that the placenta is progressively depleting maternal DHA stores during fetal growth [36]. This partially explains the progressive decline in essential fatty acids, AA and DHA maternal status towards the end of pregnancy [34,37]. Placental transfer of fatty acids involves a multi-step process of uptake and translocation facilitated by fatty acids-binding proteins. Proteins that favor ω-6 and ω-3 fatty acids over nonessential fatty acids and AA and DHA over LA and LNA have been identified [38]. Fatty acids provided by the placenta are transported to the fetal liver, esterified and secreted in lipoproteins, of which HDL is prominent. Fetal lipoprotein phospholipids, triglycerides and cholesteryl esters have AA and DHA 300- to 400-fold higher, whereas their LA and LNA precursors are lower than in infants after birth or in maternal plasma [38].

Unesterified fatty acids input into placental and mammalian cells occurs through both a passive flip-flop as well as a saturable protein-mediated mechanism by plasma membrane fatty acids-binding protein (FABPpm/GOT2), fatty acid translocase (FAT/CD36) [39], and a family of fatty acids-transport proteins (FATPs) [40]. Although the roles of these proteins in placental fatty acids uptake and metabolism are yet to be fully understood, their complex interaction has been suggested to be involved in the enrichment of LC-PUFAs that occurs in the fetal circulation compared with the maternal side. mRNA expression of FATP-1—particularly of FATP-4—in placental tissue was positively correlated with the uptake of maternal DHA into placental and cord blood phospholipids, which appears to indicate a mechanism for selective maternal/fetal DHA transfer in humans [41]. This is indicative in regard to the physiological importance of DHA because there is a preferential placental transfer, which is mediated by specific fatty acids transfer proteins (FATP-1 and FATP-4) and membrane-binding proteins that favor placental transport of DHA over other fatty acids such as LA [42,43].

Secretion of LC-PUFAs into Human Milk

Lactation has been described as the greatest physiological stress of the life cycle, largely because of the increased amounts of protein and energy needed to produce breast milk. However, among nutrition researchers, it is generally well accepted that the energy stored as fat during pregnancy can provide one-third of the energy cost of lactation during the first 3 months [44]. This fat includes LC-PUFAs. During lactation, the mother's body loses ~70–80 mg DHA/day to breast milk in addition to the amount lost to oxidation or used to fulfill the mother's own requirements. Also, in women who have resumed menstruation, DHA may also be lost in menstrual blood [27].

After birth, all of the ω-6 and ω-3 fatty acids required by the newborn must be derived from the milk or formula diet and later from complementary foods [19]. The lipids of human milk and formula are of critical importance during this time period for several reasons. First, they are a major energy source to support appropriate growth. Both human milk and infant formula provide about half their calories as fat. In addition, infants cannot synthesize enough ω-3 and ω-6 LC-PUFAs [6]. Fat is the most variable macronutrient in human milk, generally providing 40–50% of energy from fat. Milk fat consists mainly of triglycerides, representing about 98% of total milk fat. These are formed, among other fatty acids, by DHA and AA, where the fatty acids composition is also exceedingly variable. Human milk contains more than 150 different fatty acids, of which LA, LNA, AA, DHA, and several other ω-6 and ω-3 fatty acids typically make up 15 to 20% of all the fatty acids present (Table **1**) [38].

Table 1: Fatty Acids Composition (weight %) in Human Breast Milk Worldwide.

Author	Del Prado M., *et al.*	Krasevec J. M., *et al.*	Marangoni F., *et al.*	Sala-Vila A., *et al.*	**Szabo E., *et al.*	**Szabo E., *et al.*	Meneses F., *et al.*	Glew R. H., *et al.*	Karl-Göran S., *et al.*	Peng Y., *et al.*
Country	Mexico	Cuba	Italy	Spain	Germany	Turkey	Brazil	USA	Sweden	China
					SFA					
6:0	NR	NR	NR	NR	NR	NR	NR	0.017 ± 0.02	NR	NR
8:0	0.2 ± 0.05	0.17 ± 0.05	NR	NR	NR	NR	0.21 ± 0.08	0.10 ± 0.02	NR	NR
10:0	1.7 ± 0.3	1.6 ± 0.3	NR	NR	2.1 (0.02-4.8)	2.3 (0.6-3.8)	1.8 ± 0.4	1.09 ± 0.3	NR	NR
12:0	6.9 ± 1.6	7.8 ± 2.0	NR	NR	6.3 (1.9-13.9)	7.6 (2.8-13.4)	8.2 ± 2.4	4.3 ± 1.3	4.7 ± 2.5	3.04 ± 1.3
14:0	6.0 ± 1.6	9.0 ± 2.9	5.5 ± 0.4	NR	7.1 (2.7-13.0)	7.3 (3.4-11.9)	7.7 ± 2.7	5.0 ± 1.5	8.2 ± 3.0	6.2 ± 1.4
15:0	NR	NR	NR	NR	NR	NR	0.19 ± 0.1	0.28 ± 0.08	NR	NR
16:0	16.5 ± 1.4	19.4 ± 2.3	24.0 ± 0.6	21.0 ± 0.6	22.7 (18.5-30)	19.6 (12.1-24.9)	15.9 ± 2.1	18.8 ± 2.1	27.4 ± 2.1	28.5 ± 3.6
18:0	4.8 ± 0.8	4.6 ± 0.8	9.5 ± 0.8	7.6 ± 1.07	7.9 (4.1-22.5)	6.8 (4.3-13.5)	5.3 ± 1.7	6.4 ± 1.2	6.8 ± 1.0	4.8 ± 0.9
20:0	NR	NR	0.2 ± 0.01	NR	0.26 (0.02-0.8)	0.24 (0.2-0.6)	0.15 ± 0.06	0.17 ± 0.03	NR	0.13 ± 0.05
22:0	0.1 ± 0.03	NR	0.13 ± 0.02	NR	0.14 (0.01-1.1)	0.14 (0.03-0.5)	0.09 ± 0.09	0.12 ± 0.05	NR	0.11 ± 0.08
24:0	0.05 ± 0.02	NR	0.08 ± 0.01	NR	0.12 (<0.01-1.5)	0.15 (0.04-1.3)	0.11 ± 0.09	0.02 ± 0.01	NR	0.14 ± 0.05
					MUFAs					
14:1ω-5	NR	NR	NR	NR	NR	NR	0.05 ± 0.05	NR	NR	0.03 ± 0.04

Table 1: cont....

14:1ω-7	NR	NR	NR		NR	NR	0.11 ± 0.08	NR	NR	NR
16:1ω-7	2.5 ± 0.5	4.07 ± 1.04	2.3 ± 0.14	NR	2.7 (0.3-6.3)	1.7 (0.3-3.9)	1.6 ± 0.7	2.2 ± 0.6	2.7 ± 0.6	2.4 ± 0.9
18:1ω-7	2.3 ± 1.6	*	3.3 ± 0.5	NR	1.9 (0.12-11.2)	1.5 (0.6-7.1)	NR	1.52 ± 0.4	NR	NR
16:1ω-9	NR	NR	NR	NR	NR	NR	0.2 ± 0.1	0.14 ± 0.04	NR	NR
17:1ω-9	NR	NR	NR	NR	NR	NR	0.22 ± 0.6	NR	NR	NR
18:1ω-9	29.3 ± 2.6	*29.7 ± 4.1	38.9 ± 1.4	34.6 ± 1.03	30.9 (19.3-42.8)	28.0 (21.3-36.8)	24.6 ± 5.0	29.4 ± 3.4	35.9 ± 4.1	32.6 ± 3.7
19:1ω-9	NR	NR	NR	NR	NR	NR	0.13 ± 0.06	NR	NR	NR
20:1ω-9	0.3 ± 0.01	0.5 ± 0.1	0.49 ± 0.03	NR	0.35 (0.01-6.5)	0.3 (0.03-0.6)	0.27 ± 0.08	0.38 ± 0.1	NR	NR
22:1ω-9	NR	NR	0.08 ± 0.01	NR	0.06 (<0.01-0.4)	0.05 (<0.01-0.30)	0.07 ± 0.05	NR	NR	NR
24:1ω-9	NR	NR	0.12 ± 0.01	0.31 ± 0.01	0.06 (<0.01-0.8)	0.06 (<0.01-0.4)	NR	0.03 + 0.01	0.36 ± 0.2	0.45 ± 0.18
20:3ω-9	NR	NR	NR	NR	NR	NR	NR	NR	0.04 ± 0.02	0.07 ± 0.10
PUFAs ω-6										
18:2	27.3 ± 2.6	19.4 ± 4.6	12.7 ± 0.6	15.9 ± 1.4	9.8 (1.3-27.9)	15.9 (6.8-33.9)	17.3 ± 4.8	19.4 ± 4.6	9.1 ± 1.9	17.3 ± 3.4
18:3	NR	0.9 ± 0.3	0.17 ± 0.05	NR	0.11 (<0.01-2.6)	0.13 (0.03-0.5)	0.13 ± 0.09	0.12 ± 0.04	0.05 ± 0.05	0.08 ± 0.12
20:2	NR	NR	0.3 ± 0.04	0.50 ± 0.04	0.2 (0.01-0.6)	0.28 (0.06-0.8)	0.37 ± 0.1	0.60 ± 0.1	NR	1.09 ± 0.2
20:3	NR	0.5 ± 0.1	0.4 ± 0.02	0.37 ± 0.03	0.3 (0.01-2.2)	0.4 (0.07-0.9)	0.31 ± 0.1	0.37 ± 0.08	0.43 ± 0.1	0.83 ± 0.38
20:4	0.4 ± 0.1	0.7 ± 0.2	0.5 ± 0.02	0.41 ± 0.05	0.45 (0.01-1.2)	0.5 (0.04-0.9)	0.40 ± 0.15	0.44 ± 0.09	0.53 ± 0.09	0.65 ± 0.16
22:2	NR	NR	NR	0.11 ± 0.01	NR	NR	NR	0.02 ± 0.001	NR	NR
22:4	NR	0.15 ± 0.07	0.09 ± 0.01	0.02 ± 0.00	0.06 (<0.01-0.5)	0.05 (<0.01-0.5)	NR	0.10 ± 0.02	NR	NR
22:5	NR	NR	0.05 ± 0.00	NR	NR	NR	NR	0.03 ± 0.01	NR	NR
PUFAs ω-3										
18:3	1.01 ± 0.3	0.9 ± 0.2	0.7 ± 0.04	0.49 ± 0.04	0.7 (0.03-3.5)	0.7 (0.40-2.7)	1.08 ± 0.5	1.2 ± 0.4	1.31 ± 0.4	0.76 ± 0.2
18:4	NR	NR	NR	NR	NR	NR	0.32 ± 0.1	NR	NR	NR
20:3	NR	NR	NR	NR	0.04 (<0.01-0.5)	0.03 (<0.01-0.5)	0.06 ± 0.05	NR	NR	NR
20:4	NR	NR	NR	NR	NR	NR	0.09 ± 0.08	NR	NR	NR
20:5	0.5 ± 0.1	0.12 ± 0.07	0.06 ± 0.01	0.06 ± 0.01	0.05 (<0.01-0.7)	0.03 (0.01-0.6)	0.05 ± 0.01	0.08 ± 0.03	0.09 ± 0.03	0.16 ± 0.12
22:5	NR	0.15 ± 0.07	0.15 ± 0.02	0.10 ± 0.01	0.07 (<0.01-1.4)	0.04 (<0.01-0.5)	0.21 ± 0.1	0.1 ± 0.03	NR	NR
22:6	0.2 ± 0.04	0.4 ± 0.3	0.35 ± 0.06	0.18 ± 0.02	0.18 (0.01-3.1)	0.11 (0.01-2.2)	0.20 ± 0.1	0.1 ± 0.05	0.45 ± 0.2	0.61 ± 0.46

Values are reported as mean ± SD. *18:1ω-9 + 18:1ω-7. **Median (range).
SFA, saturated fatty acids; NR, not reported.

Studies concerning fatty acids composition in human breast milk worldwide revealed that women from the Dominican Republic, Japan and Canadian Arctic have the highest concentration of DHA (Table **2**) [4]. LC-PUFAs profiles of breast milk vary with nutritional habits of lactating mothers. This has been well demonstrated in human and animal studies [45]. Dietary fatty acids intakes during lactation clearly influence maternal/infant transfer of fatty acids with beneficial or adverse effects on infant development [38]. It has been observed that a diet rich in fish oil, consumption of cod liver oil or AA and DHA supplements daily (170–260 mg/day) between 2 and 8 weeks postpartum increases the content of DHA and other polyunsaturated fats in maternal plasma phospholipid and breast milk DHA concentrations and resulted in higher plasma phospholipid and plasma phospholipid DHA concentrations in infants [46–48]. It is noteworthy that DHA and AA are required for normal growth and maturation of numerous organ systems, most importantly the central nervous system (CNS) and eyes of the newborn or fetus [6].

Table 2: Worldwide concentrations of breast milk DHA and AA.

Country	Infant age	Subjects	DHA	AA
	Months	*n*	% of total fatty acids	% of total fatty acids
Pakistan	12	8	0.06	0.26
Rural South Africa	6.5	18	0.10	1.00
Canada	4	43	0.12	0.51
France	1.5	15	0.14	0.24
Netherlands	3	25	0.14	0.33
Canada	0.75–1	198	0.14	0.35
China (Enshi)	2–18	9	0.15	0.35
USA	4	29	0.15	0.48
Israel	3–10	19	0.15	0.49
Canada	1–12	48	0.17	0.37
USA	1–12	49	0.17	0.45
Finland	3	10	0.18	0.33
Finland	4	16	0.18	0.33
Sweden	6	17	0.18	0.34
China	3	23	0.18	0.51
Netherlands	3	25	0.19	0.34
Netherlands	>0.3	29	0.19	0.40
USA	2	6	0.19	0.53
USA	0.5	11	0.19	0.59
Hungary	1	18	0.19	0.59
Australia	4	33	0.20	0.39
México	5–6	10	0.20	0.40
USA	6	7	0.20	0.40
USA (Texas)	4	77	0.20	0.44
USA (Oregon)	2–11	14	0.20	0.50
Canada	3	56	0.20	0.50
Canada	>3	17	0.20	0.50
Nigeria (Niger)	0.3–6	34	0.20	0.51
Nigeria (Niger)	0.3–16	89	0.20	0.57
South Africa (Durban)	6.8	12	0.20	0.60
Australia	4	23	0.21	0.40
Australia	3	12	0.21	0.41
USA (North Carolina)	3	22	0.21	0.41
Germany	1.5	5	0.21	0.43
Belize	>0.3	6	0.21	0.44
Germany	3–4	15	0.22	0.36
China (Xichang)	2–18	10	0.22	0.52
Australia	1–12	48	0.23	0.38
United Kingdom	1–12	44	0.24	0.36
France	1	24	0.24	0.36
Sweden	3	19	0.25	0.38
Australia	3	36	0.26	0.38
Mexico	1–12	46	0.26	0.42
Australia	1	27	0.26	0.46

Table 2: cont....

Netherlands	0.5–1	5	0.26	0.47
Netherlands	0.5–1	5	0.26	0.47
Spain	3	11	0.28	0.41
Italy	6	10	0.28	0.50
China (Beijing)	2–18	10	0.28	0.63
Iceland	2	59	0.30	0.32
Canada	NR	5	0.3	0.40
Spain	0.5–1	10	0.31	0.49
France	0–3	10	0.32	0.50
France	NR	25	0.32	0.52
Panama	0.5–1	8	0.32	0.52
Nigeria	0.1–0.5	28	0.32	0.58
Nigeria	6–7	15	0.33	0.44
China	1	18	0.33	0.63
Spain	0.6–1	40	0.34	0.50
Nigeria	2–3	20	0.34	0.56
Denmark	4	39	0.35	0.30
China	1–12	50	0.35	0.49
Italy	3	73	0.35	0.50
USA (Connecticut)	0.5	5	0.37	0.67
Norway	0.75	22	0.38	0.34
France	3	28	0.38	0.50
Israel	1.5–2.5	26	0.38	0.59
Spain	0.5–1	8	0.38	0.69
Dominican Republic	0.75	7	0.40	0.50
Canada (Vancouver)	2–4	12	0.40	0.70
Suriname	>0.3	20	0.41	0.58
Chile	1–12	50	0.43	0.42
Cuba	2	52	0.43	0.67
Curacao	>0.3	47	0.43	0.71
Norway	3	111	0.47	0.37
Japan	2.3–3.3	53	0.53	0.36
Sweden	4	14	0.53	0.44
St. Lucia	1	12	0.53	0.58
Congo	5	102	0.55	0.44
Philippines	1–12	54	0.74	0.39
Dominican Republic	>0.3	6	0.91	0.33
Japan	1–12	51	0.99	0.40
Japan	0.3	20	1.10	1.00
Canadian Arctic	1–7	5	1.40	0.60

NR, not reported; DHA, docosahexaenoic acid; AA, arachidonic acid.

LC-PUFAs AND EARLY HUMAN DEVELOPMENT

LC-PUFAs in the Central Nervous System Development

The CNS has the second largest concentration of lipids after adipose tissue [49], and these lipids are an integral part of the morphology of neural cells. Saturated and monounsaturated fatty acids are components of the myelin sheaths that surround axons, and LC-PUFAs such as DHA and AA are avidly incorporated into the gray matter of the cerebral cortex. The incorporation of these fatty acids into the nerve cell membranes of the brain is one of the processes of perinatal development that contributes to the functional maturation of the CNS [50]. In vertebrates, DHA and AA contribute to the framework of the nerve cell membranes. DHA is actually enriched in brain grey matter phospholipids and represents 3–5% of the dry weight of this tissue. Phospholipids are most abundant in brain gray matter and contain high proportions of DHA in phosphatidylethanolamine (PE) and phosphatidylserine (PS) and high amounts of AA in phosphatidylinositol (PI). AA is also present in membrane phospholipids, particularly in PI throughout the body [38]. The importance of lipids for growth and development of the CNS was addressed by Widdowson in 1968 but has been given more attention during recent decades [14]. DHA and AA are rapidly incorporated in the brain during its growth spurt [38,51]. Human brain growth is at peak velocity during the last trimester of pregnancy and remains high during the first year of life with continued growth for the next several years [6]. This leads to the concept that the brain development of fetus, newborn and infant is particularly vulnerable to DHA deficiency. [38,51]. In fact, from 26 weeks of gestation until term, 80% of the brain DHA accrues in the fetus [52]. However, AA and DHA accumulation in brain continues after birth, reaching a total brain DHA deposition of about 4 g between 2 and 4 years of age [53]. DHA accumulates preferentially during the third trimester of pregnancy because the spurt of synaptogenesis and photoreceptor cell development occurs during this period [54]. Recent studies have also stressed the importance of DHA for normal development of the glial cell [55,56]. Therefore, the rapid synthesis of brain tissue with cellular differentiation and active synaptogenesis has a special need for DHA and AA [14]. It is during this period when DHA accretion into the fetal brain and nervous system is at its greatest velocity for their development [1].

Although DHA and AA are most important for the development of the CNS, AA is found in relatively large amounts in the non-neuronal membranes of all tissues, whereas DHA concentration is minor about 5-10% of total fatty acids in liver and heart membranes [57]. Therefore, the preferential incorporation of DHA in the brain as compared to other tissues is a remarkable constant throughout the evolution of the species [50]. The importance of LC-PUFAs to the CNS has been observed because dietary deficiency of ω-3 fatty acids decreases brain and retina development, impairs neurogenesis, alters gene expression and neurotransmitter including dopamine and serotonin metabolism, and decrease the kinetics of the visual photocycle [48,58]. Although AA and DHA are indispensable to brain and retina development of the fetus and newborn, currently no consensus exists whether dietary supplementation of LC-PUFAs has benefits for visual and cognitive development of infants. Eilander *et al.* evaluated the most recent evidence of available randomized controlled trials on the effect of ω-3 LC-PUFAs supplementation (with or without ω-6 LC-PUFAs) in pregnancy, lactation, infancy and childhood on visual and cognitive function during infancy and later childhood. The authors conclude that there is still limited and inconsistent evidence that supplementing mothers, infants or children with ω-3 LC-PUFAs (possibly with additional ω-6 fatty acids) can improve visual and cognitive development of infants or children. However, there are promising indications for effects of supplementing pregnant and lactating mothers with ω-3 LC-PUFAs on cognitive development of their children that warrant further studies in these and other target groups [59]. For instance, it has been proposed that maternal concentration of ω-3 LC-PUFAs during pregnancy may be important for later (7 years of age) cognitive function such as sequential processing [60].

LC-PUFAs in Photoreceptor Cells

DHA is also the major structural lipid component of retinal photoreceptor (rods and cones) outer segment membrane, where its fluidity is essential to accommodate the extremely rapid conformational changes of rhodopsin [61], and ~10% of the weight is lipid. About half of this lipid is phospholipid, with ~20% cholesterol, 15–20% cerebrosides, and smaller amounts of sulfatides and gangliosides [38]. Lipid composition predisposes the disk and plasma membranes to be fluid and structurally disordered (~84%) around physiological temperature [62]. These membranes are specialized for the rapid transmission of light and contain 90–95% of the lipid as phospholipid [38]. The fluid phospholipid environment of the disk membrane is considered to be vital for photoreceptor cell function. Phospholipid bilayers rich in DHA have higher fluidity and enhanced rates of fusion and permeability, characteristics for the normal functioning of photoreceptor cells [62]. In accordance with this, DHA concentrations in the erythrocyte phospholipids of infants have been positively associated with electroretinogram response after birth [63] and with response to visual stimulation until the age of 2 years [64]. DHA as an acyl group can constitute as much as 50% of the fatty acids in PE, PS and phosphatidylcholine (PC) and as much as 80% of all the polyunsaturated fatty acids of retinal photoreceptors [62,65]. Studies *in vivo* and *in vitro* support the idea that upon entering the photoreceptor cell, DHA is incorporated in phospholipid molecules, and certain molecular species containing DHA remain there for the life of the disk [62].

Retina and brain have a preference for DHA over ω-6 PUFAs. Each has a mechanism for conservation of DHA during fatty acids deficiency. Moreover, the retina is known to tenaciously retain its DHA, even under extremely harsh conditions of ω-3 fatty acids deficiency [62]. Unlike DHA, other ω-3 LC-PUFAs do not accumulate to any appreciable extent in the growing brain and eye [6].

LC-PUFAs in Fetal Growth

LC-PUFAs have important functions during gestation because they provide benefits for the developing fetus. It has been proposed that marine ω-3 LC-PUFAs ingested during pregnancy prolong duration of pregnancy and increase growth rate in humans. Although LC-PUFAs are considered essential for fetal growth, evidence of their growth-promoting effect is limited. Results of the few small-sized observational studies that directly measured ω-3 and ω-6 fatty acids concentrations, either maternal, neonatal [66], or both, are inconclusive [67,68]. Some evidence exists that maternal LC-PUFAs supplementation is associated with a small increment during the duration of pregnancy; however, implications of this finding for later growth and development are unclear. Further studies with larger sample sizes that take into account confounding factors are needed to examine the effects of such supplementation on growth measurements and low birth weight rates. Evidence from existing randomized clinical trials is restricted to maternal ω-3 LC-PUFAs intake only. Although positive associations are reported, these are commonly interpreted as a consequence of a prolonged gestation rather than a direct effect on fetal growth [69]. However, results of a large, community-based cohort study suggest that the maternal ω-3 and ω-6 fatty acids status early in pregnancy affects fetal growth. After adjustment for relevant covariables, both low maternal plasma concentrations of ω-3 eicosapentaenoic acids (EPA) and docosapentaenoic acid (DPA) and ω-6 LC-PUFAs dihomo-γ-linolenic acid and high concentrations of the AA were associated with birth weight and a 40–50% increase in risk of a small-for-gestational-age newborn [70]. Supporting this study, DHA and AA concentrations measured in various umbilical domains considered to reflect fetal LC-PUFAs availability during late gestation are mainly negatively related to birth weight and birth length. The observed negative relationship may result from a limited maternal/fetal LC-PUFAs transfer capacity, possibly due to low placental weight [71]. According to this, low placental weight was associated with lower plasma concentrations of AA and DHA. It has been proposed that reduced concentrations of these fatty acids were also associated with short gestation and small head circumference. AA, however, was specifically associated with weight but not gestational age. Also, a reduction in circulating AA was found in one study to be associated with reduced weight gain, which appeared to be corrected when AA concentrations were normalized [36]. In this context, the high birth weights and long duration of pregnancies observed in the fish-eating community of the Faroe Islands in the North Atlantic led to the suggestion that fatty acids from marine food may delay spontaneous delivery as well as to increase birth weight [65].

Similar findings on low AA status in preterm infants have been reported. It is thus plausible that AA, the major acyl component of inositol phosphoglycerides, is a growth factor. The probable mode of action for AA in growth may be increased blood perfusion resulting from its contribution to endothelial growth and function through the well-described breakdown of inositol phosphoglycerides or protein kinase C activation or to both processes [31]. Consequently, an inadequate LC-PUFAs supply during the neonatal period is hypothesized to contribute to this poor developmental outcome [15].

In summary, available evidence continues to suggest that fetus and neonate should receive ω-3 LC-PUFAs through placental transfer, breast milk or formula, respectively, to neural and visual system development. Availability of these fatty acids is dependent on maternal status. Therefore, education should be provided and be directed towards pregnant women, lactating women and women who intend to breastfeed regarding adequate consumption of these fatty acids. To accomplish these educational initiatives, food and nutrition professionals or other health care professionals should recommend the consumption of fish rich in DHA. Whether supplementation during pregnancy or during lactation (or both) carries more importance remains to be elucidated. It will be apparent to the reader that advances in the understanding of ω-3 LC-PUFAs for fetal growth and development and increment in pregnancy duration are insufficient because implications of this finding regarding subsequent growth and development of children are not clear. Rather, future advances will benefit from a marriage of new knowledge on the functional roles of DHA and AA with the application of sensitive tests of neural and retinal function to probe the physiologically important pools of DHA and AA in developing infants. These functional roles of LC-PUFAs include changes in the fatty acids composition of membrane. As a result, they modulate the dynamic properties of the membranes to change cellular environments with implications for cell structure and function. This exerts an influence on each of the cellular pathways during each phase of human development. Available evidence continues to suggest that fetus and neonate should receive ω-3 LC-PUFAs through placenta, breast milk or formula.

REFERENCES:

[1]   Makrides, M. Is there a dietary requirement for DHA in pregnancy?, *Prostaglandins Leukot Essent Fatty Acids.*, **2009**, *81*, 171–174.

[2]   Weseler, A. R.; Dirix, C. E. H.; Bruins, M. J.; Hornstra, G. Dietary arachidonic acid dose-dependently increases the arachidonic acid concentration in human milk. *J. Nutr.*, **2008**, *138*, 2190–2197.

[3]   Jensen, C. L.; Lapillonne, A. Docosahexaenoic acid and lactation. *Prostaglandins Leukot. Essent. Fatty Acids,* **2009**, *81*, 175–178.

[4]   Brenna, J. T.; Varamini, B.; Jensen, R. G.; Diersen-Schade, D. A.; Boettcher, J. A.; Arterburn, L. M. Docosahexaenoic and arachidonic acid concentrations in human breast milk worldwide. *Am. J. Clin. Nutr.*, **2007**, *85*, 1457–1464.

[5]   Innis, S. M. Omega-3 fatty acids and neural development to 2 years of age: do we know enough for dietary recommendations?. *J. Pediatr. Gastroenterol. Nutr.,* **2009**, *48*, S16–S24.

[6]   Koletzko, B.; Lien, E.; Agostoni, C.; Bolees, H.; Campoy, C.; Cetin, I.; Decsi, T.; Dudenhausen, J. W.; Dupont, C., Forsyth, S.; Hoesli, I.; Holzgreve, W.; Lapillonne, A.; Putet, G.; Secher, N. J.; Symonds, M.; Szajewska, H.; Willatts, P.; Uauy, R. The roles of long-chain polyunsaturated fatty acids in pregnancy, lactation and infancy: review of current knowledge and consensus recommendations. *J. Perinat. Med,.* 2008, *36*, 5-14.

[7]   Helland, I. B.; Smith, L.; Saarem, K.; Saugstad, O. D.; Drevon, C. A. Maternal supplementation with very-long-chain n-3 fatty acids during pregnancy and lactation augments children's IQ at 4 years of age. *Pediatrics,* **2003**, *111*, e39-e44.

[8]   Knopp, R. H. Hormonal-mediated changes in nutrient metabolism in pregnancy: a physiological basis for normal fetal development. *Ann. N. Y. Acad. Sci.,* **1997**, *817*, 251-271.

[9]   Herrera, E. Metabolic adaptations in pregnancy and their implications for the availability of substrates to the fetus. *Eur. J. Clin. Nutr.,* **2000**, *54*, S47-S51.

[10]   Haggarty, P. Placental regulation of fatty acid delivery and its effect on fetal growth—a review. *Placenta,* **2002**, *23*, S28-S38.

[11]   Otto, S. J.; van Houwelingen, A. C.; Badart-Smook, A.; Hornstra, G. Comparison of the peripartum and postpartum phospholipids polyunsaturated fatty acid profiles of lactating and nonlactating women. *Am. J. Clin. Nutr.,* **2001**, *73*, 1074-1079.

[12]   Torres, A. G.; Trugo, N. M. F. Evidence of inadequate docosohexaenoic acid status in Brazilian pregnant and lactating women. *Rev. Saúde Pública,* **2009**, *43*, 359-368.

[13]   Joardar, A:, Sen, A. K.; Das, S. Docosahexaenoic acid facilitates cell maturation and beta-adrenergic transmission in astrocytes. *J. Lipid Res.,* **2006**, *47*, 571-581.

[14]   Sabel, K. G.; Lundqvist-Persson, C.; Bona, E.; Petzold, M.; Strandvik, B. Fatty acid patterns early after premature birth, simultaneously analysed in mothers' food, breast milk serum phospholipids of mothers and infants. *Lip. Health Dis.,* **2009**, *8*, 20-35.

[15]   Makrides, M.; Gibson, R. A.; McPhee, A. J.; Collins, C. T., Davis, P. G.; Doyle, L. W.; Simmer, K.; Colditz, P. B.; Morris, S.; Smithers, L. G.; Willson, K.; Ryan, P. Neurodevelopmental outcomes of preterm infants fed high-dose docosahexaenoic acid: a randomized controlled trial. *JAMA,* **2009**, *301*, 175-182.

[16]   Al, M. D.; van Houwelingen, A. C.; Kester, A. D.; Hasaart. T. H.; de Jong, A. E.; Hornstra, G. Maternal essential fatty acid patterns during normal pregnancy and their relationship to the neonatal essential fatty acid status. *Br. J. Nutr.,* **1995**, *74*, 55-68.

[17]   Herrera, E. 'Implications of dietary fatty acids during pregnancy on placental, fetal and postnatal development—a review', *Placenta,* **2002**, *23*, S9-19.

[18]   Uauy, R.; Mena, P.; Rojas, C. Essential fatty acids in early life: structural and functional role. *Proc. Nutr. Soc.,* **2000**, *59*, 3–15.

[19]   Rodriguez, A.; Sarda, P.; Nessman, C.; Boulet, P.; Leger, C. L.; Descomps, B. D6- and D5-Desaturase activities in the human fetal liver: kinetic aspects. *J. Lipid Res.,* **1998**, *39*, 1825-1832.

[20]   Uauy, R.; Mena, P.; Wegher, B.; Nieto, S.; Salem, N. Long chain polyunsaturated fatty acid formation in neonates: effect of gestational age and intrauterine growth. *Pediatr. Res.,* **2000**, *47*, 127-135.

[21]   Vlaardingerbroek, H.; Hornstra, G. Essential fatty acids in erythrocyte phospholipids during pregnancy and at delivery in mothers and their neonates: comparison with plasma phospholipids. *Prostaglandins Leukot. Essent. Fatty Acids,* **2004**, *71*, 363–374.

[22]   Herrera, E. Implications of dietary fatty acids during pregnancy on placental, fetal and postnatal development—a review. *Placenta*, **2002**, *23*, S9–S19.

[23]   Rodríguez-Cruz, M.; Tovar, A. R.; Palacios-González, B.; Del Prado, M.; Torres, N. Synthesis of long-chain polyunsaturated fatty acids in lactating mammary gland: role of Delta5 and Delta6 desaturases, SREBP-1, PPARalpha, and PGC-1. *J. Lipid Res.,* **2006**, *47*, 553-560.

[24]   Demmelmair, H.; Baumheuer, M.; Koletzko, B.; Dokoupil, K.; Kratl, G. Metabolism of U-13C-labeled linoleic acid in lactating women. *J. Lipid Res.*, **1998**, *39*, 1389–1396.

[25]   Del Prado, M.; Villalpando, S.; Elizondo, A.; Rodríguez, M.; Demmelmair, H.; Koletzko, B. Contribution of dietary and newly formed arachidonic acid to human milk lipids in women eating a low-fat diet. *Am. J. Clin. Nutr.,* **2001**, *74*, 242-247.

[26]   Makrides, M. Outcomes for mothers and their babies: do n-3 long-chain polyunsaturated fatty acids and seafoods make a difference?. *J. Am. Diet. Assoc.*, **2008**, *108*, 1622-1626.

[27]   Makrides, M.; Gibson, R. A. Long-chain polyunsaturated fatty acid requirements during pregnancy and lactation. *Am. J. Clin. Nutr.,* **2000**, *71*, 307S–311S.

[28]   Levant, B.; Ozias, M. K.; Carlson, S. E.; Diet (n-3) polyunsaturated fatty acid content and parity interact to alter maternal rat brain phospholipid fatty acid composition. *J. Nutr.,* **2006**, *136*, 2236–2242.

[29]   Del Prado, M.; Villalpando, S.; Gordillo, J.; Hernández-Montes, H. A high dietary lipid intake during pregnancy and lactation enhances mammary gland lipid uptake and lipoprotein lipase activity in rats. *J. Nutr.,* **1999**, *129*, 1574–1578.

[30]   Rodríguez-Cruz, M.; Sánchez, R.; Maldonado, J.; Bernabe, M.; Del Prado, M.; López-Alarcón, M. Effect of dietary levels of corn oil on maternal arachidonic acid synthesis and fatty acid composition in lactating rats. *Nutrition*, **2009**, *25*, 209-215.

[31]   Fidler, N.; Sauerwald, T.; Pohl, A.; Demmelmair, H.; Koletzko, B. Docosahexaenoic acid transfer into human milk after dietary supplementation: a randomised trial. *J. Lipid Res.,* **2000**, *41*, 1376–1383.

[32]   Nelson, G. J.; Schmidt, P. C.; Bartolini, G.; Kelley, D. S.; Phinney, S. D.; Kyle, D.; Silbermann, S.; Schaefer, E. J. The effect of dietary arachidonic acid on plasma lipoprotein distributions, apoproteins, blood lipid levels, and tissue fatty acid composition in humans. *Lipids,* **1997**, *32*, 427–433.

[33]   Calder, P. C. Dietary arachidonic acid: harmful, harmless or helpful?. *Br. J. Nutr.,* **2007**, *98*, 451–453.

[34]   Uauy, R.; Dangour, A. D. Nutrition in brain development and aging role of essential fatty acids. *Nutr. Rev.* **2006**, *64*, S24-S33.

[35]   Elias, S. L.; Innis, S. M. Infant plasma trans, n-6, and n-3 fatty acids and conjugated linoleic acids are related to maternal plasma fatty acids, length of gestation, and birth weight and length. *Am. J. Clin. Nutr.*, **2001**, *73*, 807-814.

[36]   Crawford, M. A. Placental delivery of arachidonic and docosahexaenoic acids: implications for the lipid nutrition of preterm infants. *Am. J. Clin. Nutr.,* **2000**, 71, 275S–284S.

[37]   Al, M. D. M.; Van Houwelingen, A. C.; Hornstra, G. Long chain polyunsaturated fatty acids, pregnancy, and pregnancy outcome. *Am. J. Clin. Nutr.,* **2000**, *71*, S285- S291.

[38]   Innis, S. M. Fatty acids and early human development. *Early Hum. Dev.,* **2007**, 83, 761–766.

[39]   Thomas, B.; Ghebremeskel, K.; Lowy, C.; Min, Y.; Crawford, M. A. Plasma AA and DHA levels are not compromised in newly diagnosed gestational diabetic women. *Eur. J. Clin. Nutr.*, **2004**, *58*, 1492–1497.

[40]   Dutta-Roy, A. K. Transport mechanisms for long-chain polyunsaturated fatty acids in the human placenta. *Am. J. Clin. Nutr.,* **2000**, *71*, 315–322S.

[41]   Larqué, E.; Krauss-Etschmann, S.; Campoy, C.; Hartl, D.; Linde, J.; Klingler, M.; Demmelmair, H.; Caño, A.; Gil, A.; Bondy, B.; Koletzko, B. Docosahexaenoic acid supply in pregnancy affects placental expression of fatty acid transport proteins. *Am. J. Clin. Nutr.,* **2006**, *84*, 853– 861.

[42]   Koletzko, B.; Larque, E.; Demmelmair, H. Placental transfer of long-chain polyunsaturated fatty acids (LC-PUFA). *J. Perinat. Med.,* **2007**, *35*, S5–11.

[43]   Larqué, E.; Demmelmair, H.; Klingler, M.; De Jonge, S.; Bondy, B.; Koletzko, B. Expression pattern of fatty acid transport protein-1 (FATP-1), FATP-4 and heart-fatty acid binding protein (H-FABP) genes in human term placenta. *Early Hum. Dev.*, **2006**, *82*, 697–701.

[44]   Taylor, K. B.; Anthony, L. E. *Nutritional aspects of pregnancy, lactation, infancy and childhood, adolescence, middle age, and old age,* McGraw-Hill: New York, **1983**.

[45]   Schaeffer, L.; Gohlke, H.; Muller, M.; Heid, I. M.; Palmer, L. J.; Kompauer, I.; Demmelmair, H.; IIlig, T.; Koletzko, B.; Henrich, J. Common genetic variants of the FADS1 and FADS2 gene cluster and their reconstructed haplotypes are associated with the fatty acid composition in phospholipids. *Hum. Mol. Genet.* **2006**, *15*, 1754–1756.

[46]   Hale, T. W.; Hartmann, P. *Textbook on Human Lactation,* Hale Publishing LP: Amarillo, TX, **2007**.

[47]   van Goor, S. A.; Dijck-Brouwer, D. A. J.; Hadders-Algra, M.; Doornbos, B.; Erwich, J. J. H.; Schaafsma, A.; Muskiet, F. A. J. Human milk arachidonic acid and docosahexaenoic acid contents increase following supplementation during pregnancy and lactation. *Prostaglandins Leukotr. Essent. Fatty Acids,* **2009**, *80*, 65–69.

[48]   Olafsdottir, A. S.; Thorsdottir, I.; Wagner, K. H.; Elmadfa, I. Polyunsaturated fatty acids in the diet and breast milk of lactating Icelandic women with traditional fish and cod liver oil consumption. *Am. Nutr. Metab.* **2006**, *50*, 270-276.

[49]   Sprecher, H. An update on the pathways of polyunsaturated fatty acid metabolism. *Curr. Opin. Clin. Nutr. Metab. Care*, **1999**, *2*, 135–138.

[50]   Jean-Marc, A.; Guesnet, P.; Vancassel, S.; Astorg, P.; Denis, I.; Langeliera, B.; Sabah, A.; Poumès-Ballihauta, C.; Champeil-Potokara, G.; Laviallea, M. Polyunsaturated fatty acids in the central nervous system: evolution of concepts and nutritional implications throughout life. *Reprod. Nutr. Dev.*, **2004**, *44*, 509–538.

[51]   Georgieff, M. K.; Innis, S. M. Controversial nutrients that potentially affect preterm neurodevelopment: essential fatty acids and iron. *Pediatr. Res.*, **2005**, *57*, 99R–103R.

[52]   Clandinin, M. T.; Chappell, J. E.; Leong, S.; Heim, T.; Swyer, P. R.; Chance, G. W. Intrauterine fatty acid accretion rates in human brain: implications for fatty acid requirements. *Early Hum. Dev.*, **1980**, *4*, 121-129.

[53]   Martinez, M. Polyunsaturated fatty acids in the developing human brain, red cells and plasma: influence of nutrition and peroxisomal disease. *J. World Rev. Nutr. Diet.*, **1994**, *75*, 70-78.

[54]   Jacobson, J. L.; Jacobson, S. W.; Muckle G.; Kaplan-Estrin, M.; Ayotte, P.; Dewailly, E. Beneficial effects of a polyunsaturated fatty acid on infant development: evidence from the Inuit of Arctic Québec. *J. Pediatr.* **2008**, *152*, 356-364.

[55]   Joardar, A.; Sen, A. K.; Das, S. Docosahexaenoic acid facilitates cell maturation and beta-adrenergic transmission in astrocytes. *J. Lipid Res.*, **2006**, *47*, 571-581.

[56]   Champeil-Potokar, G.; Chaumontet, C.; Guesnet, P.; Lavialle, M.; Denis, I., Docosahexaenoic acid (22:6n-3) enrichment of membrana phospholipids increases gap junction coupling capacity in cultured astrocytes. *Eur. J. Neurosci.*, **2006**, *24*, 3084-3090.

[57]   Bazan, N. G. *Nutrition and the brain.* New York: Raven Press Ltd., New York, **1990**.

[58]   Innis, S. M. Dietary n-3 fatty acids and brain development. *J. Nutr.* **2007**, *137*, 855–859.

[59]   Eilander, A.; Hundscheid, D. C.; Osendarp, S. J.; Transler, C.; Zock, P. L. Effects of n-3 long chain polyunsaturated fatty acid supplementation on visual and cognitive development throughout childhood: a review of human studies. *Prostaglandins Leukot. Essent. Fatty Acids,* **2007**, *76*, 189–203.

[60]   Helland, I. B.; Smith, L.; Blomén, B.; Saarem, K.; Saugstad, O. D.; Drevon, C. A. Effect of supplementing pregnant and lactating mothers with n-3 very-long-chain fatty acids on children's IQ and body mass index at 7 years of age. *Pediatrics,* **2008**, *122*, e472-e479.

[61]   Kurlak, L. O.; Stephenson, T. J. Plausible explanations for effects of long chain polyunsaturated fatty acids (LCPUFA) on neonates. *Arch. Dis. Child. Fetal Neonatal Ed*, **1999**, *80*, F148–F154.

[62]   Giusto, N. M.; Pasquare, S. J.; Salvador, G. A.; Castagnet, P. I.; Roque, M. E.; Ilincheta de Boschero, M. G. Lipid metabolism in vertebrate retinal rod outer segments. *Prog. Lipid Res.*, **2000**, *39*, 315-391.

[63]   Malcolm, C. A.; Hamilton, R.; McCulloch, D. L.; Montgomery, C.; Weaver, L. T., Scotopic electroretinogram in term infants born of mothers supplemented with docosahexaenoic acid during pregnancy. *Invest. Ophthalmol. Vis. Sci.,* **2003**, *44*, 3685–3691.

[64]   Makrides, M.; Gibson, R. A. The role of fats in the lifecycle stages Pregnancy and the first year of life. *Med. J. Australia*, **2002**, *176*, S111-S112.

[65]   Stillwell, W.; Wassall, S. R. Docosahexaenoic acid: membrane properties of a unique fatty acid. *Chem. Phys. Lipids*, **2003**, *126*, 1-27.

[66]   Lucas, M.; Dewailly, E.; Muckle, G.; Ayotte, P.; Bruneau, S.; Gingras, S.; Rhainds, M.; Holub, B. J. Gestational age and birth weight in relation to n-3 fatty acids among Inuit (Canada). *Lipids,* **2004**, *39*, 617–626.

[67]   Grandjean, P.; Bjerve, K. S.; Weihe, P.; Steuerwald, U. Birthweight in a fishing community: significance of essential fatty acids and marine food contaminants. *Int. J. Epidemiol.,* **2001**, *30*, 272– 1278.

[68]   Rump, P.; Mensink, R. P.; Kester, A. D. M.; Hornstra, G. Essential fatty acid composition of plasma phospholipids and birth weight: a study in term neonates. *Am. J. Clin. Nutr.*, **2001**, *73*, 797– 806.

[69]   Szajewska, H.; Horvath, A.; Koletzko, B. Effect of n-3 long-chain polyunsaturated fatty acid supplementation of women with low-risk pregnancies on pregnancy outcomes and growth measures at birth: a meta-analysis of randomized controlled trials. *Am. J. Clin. Nutr.,* **2006**, *83*, 1337–1344.

[70]   van Eijsden, M.; Hornstra, G.; van der Wal, M. F.; Vrijkotte, T. G. M.; Bonsel, G. J., Maternal n-3, n-6, and trans fatty acid profile early in pregnancy and term birth weight: a prospective cohort study. *Am. J. Clin. Nutr.,* **2008**, *87*, 887–895.

[71]   Dirix, C. E. H.; Kester, A. D.; Hornstra, G. Associations between neonatal birth dimensions and maternal essential and trans fatty acid contents during pregnancy and at delivery. *Br. J. Nutr.*, **2009**, *101*, 399-407.

CHAPTER 3

The Role of n-3 Long Chain Polyunsaturated Fatty Acids in Cognitive Function

Chih-Chiang Chiu, M.D.[1,2,3]*, **Robert J Stewart, M.D.**[3] **and Shih-Yi Huang PhD.**[4]

[1]*Department of Psychiatry, Taipei City Psychiatric Center, Taipei City Hospital, Taipei, Taiwan.* [2]*Department of Psychiatry, School of Medicine, Taipei Medical University, Taipei, Taiwan.* [3]*King's College London (Institute of Psychiatry), London, United Kingdom.* [4]*School of Nutrition and Health Sciences, Taipei Medical University, Taipei, Taiwan*

Abstract: Recently, there has been an increase in research into the relationship between long-chain n-3 polyunsaturated fatty acids (n-3 PUFAs) and cognitive function although the topic remains controversial. In epidemiological studies, fatty fish or n-3 PUFAs consumption has been found to be associated with a reduced risk of cognitive decline or dementia. In addition, higher proportion of total n-3 PUFAs or higher ratio of n-3 to n-6 PUFAs on erythrocyte membranes have been found to be associated with a lower risk of cognitive decline. Furthermore, in some animal models, docosahexaenoic acid (DHA, C22:6n-3) or eicosapentaenoic acid (EPA, C22:5n-3) administration have been found to improve learning ability and reduce Aβ amyloid deposition. However, findings from clinical trials of n-3 PUFAs supplementation to improve cognition in older people have been inconsistent. Some sub-group analyses have suggested that people with mild dementia or mild cognitive impairment may benefit most. Contradictory results may be clarified by better controlling possible confounders, more consistency in interventions and outcome measurements, and longer follow-up. Larger but possibly more focused trials should be considered along with efforts to develop better biomarkers for intervention response.

INTRODUCTION

As national population ages and survival rates increase, the impact of dementia and cognitive impairment in older people is becoming more significant, although still underappreciated and under-researched. Globally, close to 24.3 million people have dementia today, with 4.6 million new cases of dementia every year [1]; however, treatment and prevention of dementia remain largely neglected topics [2]. Cognitive impairment in older people not only affects their own quality of life but also that of family and other caregivers. The efficacy of pharmacotherapeutic interventions for primary and secondary prevention are still controversial [3], and increasing research has instead focused on modifying lifestyle and nutrition as alternative strategies. Within this field, fish consumption and/or n-3 polyunsaturated fatty acids (PUFAs) have attracted particular attention recently.

PUFAs include the family of omega-6(n-3) and omega-3(n-3) fatty acids. Because many of these essential fatty acids cannot be synthesized in the human body sufficiently, they must be derived from dietary sources to meet physiological requirements. Some omega-6 fatty acids, such as arachidonic acid (AA, C20:4n-3), can be manufactured in the body by using linoleic acid (LA, 18:2n-6) as a precursor. The principle n-3 PUFAs metabolites of α-linolenic acid (ALA, 18:3n-3) are eicosapentaenoic acid (EPA, C20:5n-3) and docosahexaenoic acid (DHA, C22:6n-3) (Yehuda *et al.*, 2002). However, DHA synthesis from dietary ALA is limited in humans [4]. Pre-formed DHA and EPA may instead be obtained directly from the diet, especially fatty fish, such as trout, salmon and tuna [5]. Since the brain contains very high amounts of PUFAs (constituting approximately 20 percent of its dry weight), particularly DHA and AA, associations with cognitive function and risk of dementia have been hypothesized and investigated, although results have not been consistent. Here, we first present the evidence from epidemiological research, studies of biological markers, animal models, and clinical trials to date, and then we will discuss the possible mechanisms involved and unresolved issues.

*Address correspondence to Dr. Chih-Chiang Chiu No. 309: Songde Road, Taipei 110, Taiwan. Department of Psychiatry, Taipei City Psychiatric Center, Taipei City Hospital, Taipei, Taiwan Telephone number: 886-2-27263141ext 1346 Fax number: 886-2-27272150 E-mail: DAF50@tpech.gov.tw

Maricela Rodríguez-Cruz and Mardia López-Alarcón (Eds)

DEMENTIA, ALZHEIMER'S DISEASE AND COGNITIVE IMPAIRMENT

A large number of neurodegenerative disorders are associated with the development of dementia (defined as a progressive global deterioration of higher brain functions). Alzheimer's disease (AD), the most common form of late-onset dementia, is a progressive neurodegenerative disorder manifesting with cognitive and memory deterioration, impairment of activities of daily living, and associated behavioral disturbances and neuropsychiatric symptoms [6]. Other common causes of dementia in late life include vascular dementia, dementia in Parkinson's disease, dementia with Lewy bodies, and frontotemporal dementia. In older age groups, these disorders frequently manifest in mixed, overlapping forms rather than as single, discrete entities. In most clinical and epidemiological research they are defined as 'probable' disorders based on clinical and imaging information obtained in life since, to date, it is not possible to determine the underlying pathology until *post mortem*.

The prevalence of clinical AD increases with age, present in about 5% of people aged 65 to 74 years, in almost 20% of those aged 75 to 84 years, and in more than 45% of those aged 85 years or older [7]. A pathological hallmark of AD is accumulation of beta-amyloid (A β) peptide in the brain [8]. The apolipoprotein E ε 4 allele (APOE ε 4) is associated with accelerated deposition of amyloid and higher risk of AD [9]. Neurofibrillary tangles, another pathological characteristic of AD, are aggregates of abnormally hyperphosphorylated tau protein and their presence correlates with the severity of clinical symptoms in AD [10].

To date, treatments for people with AD have been limited. Four acetyl cholinesterase inhibitors are available for patients with mild to moderate AD, while one N-methyl-D aspartate (NMDA) antagonist has been approved by the US Food and Drug Administration for people with moderate to severe AD [11]. In addition, American Psychiatric Association guidelines currently recommend against the use of vitamin E, estrogen replacement, and anti-inflammatory drugs (NSAIDs) in AD, and do not currently support the use of ginkgo biloba, selegiline, or chelating agents. Antipsychotic and other sedative agents are used sometimes for behavioral symptoms associated with dementia, but it has been recommended that these are limited because they may increase mortality [3]. Instead, non-pharmacologic interventions are recommended as first-line interventions for behavioral disturbance, and psychosocial intervention and caregiver education and support should be considered as a more holistic approach.

Mild cognitive impairment (MCI) is conceptualized as a transitional state between the cognitive changes seen associated with 'normal aging' and those seen in early dementia, being defined as cognitive decline greater than expected for an individual's age and educational level but not interfering notably with activities of daily life. Prevalence ranges from 3% to 19% in adults older than 65 years in population-based study. Some people seem to remain stable over time, but more than half progress to dementia within 5 years [12]. Identifying MCI has been recommended as a useful potential means of predicting AD and allowing earlier intervention.

N-3 POLYUNSATURATED FATTY ACIDS AND AGING

Mammalian brain tissue is predominantly composed of lipids with different profiles of saturated and monounsaturated fatty acids and PUFAs. The principal n-3 PUFA found in the human brain is DHA, comprising 10-20% of total fatty acids, whereas the other n-3 PUFAs comprise less than 1% of total fatty acids [13]. During the third trimester of pregnancy, DHA accumulates in fetus brain tissue at a rapid rate and represents approximately 9% of total cortical fatty acid composition, increasing between birth and age 20 to compose nearly 15% of total cortical fatty acids as found in *post mortem* brain tissue [14;15]. Numerous studies have found positive correlations between blood DHA levels and development of cognitive and/or visual function in breast- and formula-fed infants [16]; however, a recent Cochrane's review concluded that supplementing term formulas with PUFAs has not been proven to result in benefits with regard to cognition [17].

During aging, there are many biochemical changes in the brain affecting the neuronal membrane, whose functions include the conduction of neuronal information along the axon, regulation of membrane bound enzymes and receptors, and control of the ionic channel structures and activities [18]. Most aging studies suggest a significant decrease in the level of PUFAs in the brain [19-21], which may due to the reduction of delta-6 desaturase, enzyme desaturation of LA and ALA and hence for their further metabolism, less food intake and absorption activities [22].

In addition, the level of fatty acids incorporation into the membrane is inhibited and the turnover rate is low [23]. Combinations of these changes in aging may result in a decreased membrane fluidity, which may further affect the normal physiological functioning of the neural membrane since this is highly dependents on its structure [21].

ASSOCIATIONS OF FISH AND/OR N-3 PUFAS CONSUMPTION WITH COGNITIVE DYSFUNCTION OR DEMENTIA

Cross-sectional and prospective research investigating associations of dietary fatty acids with age-related cognitive decline, cognitive impairment, dementia and Alzheimer's disease are accumulating (Table 1). In a cross-sectional study of a community older population in the Netherlands, dietary cholesterol and to a lesser extent saturated fatty acids intake were found to be associated with cognitive impairment [24], compatible with a previous report [25], whereas fatty fish and total n-3 PUFAs (EPA+DHA) consumption were associated with a decreased risk of cognitive impairment [24]. People with a mean daily intake of fish and fish products of $\geqq$ 10 gm/day were found to have higher mean cognitive scores in the Hordaland Health Study, and the positive association between total intake of seafood and cognitive function showed dose-dependence [26]. The dose-dependent pattern is further supported by a recent cross-national report [27].

Table 1: Summarising reports of association between fish and/or n-3 PUFAs consumption and cognitive dysfunction or dementia.

	Study design	Sample	Number	Fish/or PUFAs measurement	Cognitive domains or other measurements	Principal findings
Kalmijn *et al.* 2004 [24]	Cross-sectional design	Age 45 to 70 y (mean 56.3y)	1,613	Intake of fatty fish, total fat, PUFAs, LA, ALA, DHA, EPA (FFQ)	Cognitive impairment: the lowest 10% of the compound Z-score	High intake of cholesterol and saturated fatty acids, and low fatty fish and n-3 PUFAs associates with the risk of cognitive impairment.
Nurk *et al.* 2007 (Hordaland Health Study, Norway) [26]	Cross-sectional design	Age 70-74 y	2,031	Fish and fish product, and seafood consumption (FFQ)	Poor cognitive performance: below 10 % of the cognitive test except TMT-A	Subjects with mean daily intake of fish and fish products $\geqq$ 10gm/day had a better cognitive scores and lower prevalence of poor cognition.
Albanese *et al.* 2009 (10/66 Dementia Research Group) [27]	Cross-sectional design, 11 sites of Peru, Mexico, China, and India	Age $\geqq$ 65 y	14,960	Frequency of fish intake(face-to-face interview of diet habit)	Dementia(DSM-IV)	A consistent dose-dependent inverse association between fish intake and dementia across all sites except India.
Kalmijn *et al.* 1997 (Zutphen Elderly Study, Netherlands) [28]	Cross-sectional and prospective design for 3-y follow-up	Age 69-89 y	476 in cross-section, 342 in cohort	Intake of fish, LA, n-3 PUFAs (Cross-check dietary history method)	MMSE decline in 3 years	High LA associates with cognitive impairment. High fish intake tends to inversely associate with cognitive impairment and cognitive decline.
Morris *et al.* 2005 (Chicago Health and Aging Project, USA) [29]	Prospective design for 6-y follow-up	Age $\geqq$ 65y	3,718	Fish consumption (FFQ)	Average of Z scores for 4 cognitive tests	Fish intake associates with a slower rate of cognitive decline.
van Gelder *et al.* 2007 (Zutphen Elderly Study, Netherlands) [30]	Prospective design for 5-y follow-up	Men, age70-89 y	228	Fish and EPA + DHA intake (Cross-check dietary history method)	Changes of MMSE scores	Fish consumers had significantly less cognitive decline. A linear trend for the relationship between EPA + DHA and cognitive decline.
Beydoun *et al.* 2007 (Atherosclerosis Risk In Community study, USA) [32]	Prospective design for mean follow-up of 6 y	Age 50-65 y	Diet: 7,814; Plasma levels of fatty acids: 2,251	LA, ALA, n-3 and n-6 PUFAs (FFQ and % of all plasma levels)	Cut-off points used for decline in each of three cognitive tests and composite score	Increase in dietary n-3 PUFAs and balancing n-3/n-6 PUFAs may decrease the risk of cognitive decline in verbal fluency among subjects with hypertension and hyperlipidemia

Table 1: cont....

Vercambre *et al.* 2009 (Education National Study, France) [31]	Prospective design for 13-y follow-up	Women, mean age 65.5 y.	4,758	Fish consumption (Diet history questionnaire)	Cognitive decline (DECO) and everyday functioning (IADL) by informant report.	Lower intake of poultry, fish and animal fats associate with a recent cognitive decline.
Kalmijn *et al.* 1997 (Rotterdam Study, Netherlands) [33]	Prospective design for 2.1-year follow-up	Age≧55 y, no dementia	5,386	Fish consumption, total fat and saturated fat (FFQ)	Dementia(DSM-IV), AD(NINDCS-ADRDA), and VaD (NINDCS -ADDTC)	High saturated fat and cholesterol intake increases the risk of dementia. Fish consumption may decrease this risk.
Barberger-Gateau *et al.* 2002 (PAQUID project, France) [34]	Prospective design for 7-year follow-up	Age≧68 y, no dementia	1,674	Fish, seafood, and meat consumption (FFQ)	Dementia (DSM-III-R)	Intake of fish or seafood≧once a week may lower the risk of dementia. Significance of the association decreases after adjusting education.
Engelhart *et al.* 2002 (Rotterdam Study, Netherlands) [38]	Prospective design for follow-up of 6 years.	Mean age 68(8) y, no dementia	5,395	Fish intake, intake of different kinds of fatty acids (FFQ)	Dementia(DSM-IV), AD(NINDCS-ADRDA), and VaD (NINDCS -ADDTC)	No any association of increased risk of dementia with intake of total, saturated, trans fat, cholesterol, PUFAs, or n-3PUFAs.
Morris *et al.* 2003 (Chicago Health and Aging Project) [35]	Prospective design for 3.9y follow-up	Age ≧ 65y, no dementia	815	Weekly fish consumption, n-3 PUFAs intake (self-administered FFQ)	AD(NINDCS-ADRDA)	Fish consumers≧1 /week with 60% less risk of AD compared to rare or non fish consumers. N-3 PUFAs and DHA intake, but not EPA intake, associate with reduced risk of AD.
Huang *et al.* 2005 (Cardiovascular Health Cognition Study, USA) [36]	Prospective design for mean follow-up of 5.4 years.	Age ≧ 65, no dementia.	2,233	Fatty fish or lean fish (FFQ)	Dementia(DSM-IV), AD(NINDCS -ADRDA), VaD(NINDCS –ADDTC)	Consumption of fatty fish but not lean fish associates with a reduced risk of dementia and AD for APOE ε 4 non-carriers.
Barberger-Gateau *et al.* 2007 (Three City study, France) [37]	Prospective design for mean 3.84-y follow-up	Age ≧ 65, no dementia.	8,085	Weekly fish consumers(FFQ)	Dementia(DSM-IV), and AD(NINDCS -ADRDA)	Weekly fish intake associates with a reduced risk of AD and dementia but only among APOE ε 4 non-carriers.
Devore *et al.* 2009 (Rotterdam Study, Netherlands) [39]	Prospective design for 9.6 yr follow-up	Age ≧ 55y, no dementia.	5,395	Fish intake, different fish types, n-3 PUFAs intake(FFQ)	Dementia(DSM-IV), AD(NINDCS-ADRDA), VaD(NINDCS -ADDTC)	Total fish or n-3 PUFAs intake is not associated with long-term dementia or AD risk.

ALA, α-linolenic acid; FFQ, food frequency questionnaire; DECO, observed cognitive deterioration scale; DSM, Diagnostic and Statistical Manual of Mental Disorders; NINDCS –ADDTC, National Institute of Neurological and Communication Disorders and Stroke-Association International pour la Recherche et l' Enseignement en Neurosciences; NINDCS- ADRDA, National Institute of Neurological and Communication Disorders and Stroke- Alzheimer's Disease and Related Disorder; IADL, instrumental activities of daily living; LA, linoleic acid; VaD, vascular dementia; TMT-A, Trail Making Test, part A.

Considering prospective research, high fish consumption was inversely associated with cognitive decline in the Zutphen Elderly Study [28]. Further studies have found that fish consumption [29-31], total intake of n-3 PUFAs [24;30;32], LA intake [28], and balancing intake of n-3/n-6 PUFAs [32] are inversely related to the risk of cognitive decline. Compared with cognitive decline in people who consumed fish less than weekly, the gradient was 10% slower among those who reported consuming 1 fish meal per week, and 13% slower among those who reported consuming two or more fish meals per weeks [29]. A dose-response effect was also found for the negative association between combined intake of EPA and DHA and cognitive decline [30]. In the Atherosclerosis Risk In Communities study, an interaction between hypertensive status was found, the negative association between dietary n-3 fatty acids intake and cognitive decline being stronger in people with hypertension and hyperlipidemia [32]. On the other hand, higher intake of saturated fat has been found to be positively associated with cognition decline [24;29].

Considering dementia as an outcome, a high saturated fat and cholesterol intake has been found to be associated with an increased risk of dementia, but fish consumption was associated with a reduced risk in the Rotterdam study [33]. In addition, people who ate fish or seafood at least once a week were found to have a significantly lower risk of having dementia or AD [34;35]. Similar negative associations were found for AD risk and total intake of n-3 PUFAs and DHA, but not for EPA [35]. However, this negative correlation may be weaker after adjustment for educational measures, suggesting at least partial confounding [34]. In the Cardiovascular Health Cognition Study, consumption of lean fried fish had no protective effects, whereas consumption of fatty fish was associated with a reduced risk of dementia and AD for those without the APOE ε 4 allele [36] though again adjustment by education and income attenuated this. In the Three-City study in France, only among APOE ε 4 non-carriers, weekly consumption of fish was associated with a reduced risk of AD and all causes of dementia and regular consumption of n-6 PUFAs-rich oils not compensated by consumption of n-3 PUFAs-rich oil or fish was associated with an increased risk of dementia [37]. Nevertheless, when the data of the Rotterdam study were re-examined, there was no single fat measure associated with increased risk of dementia or its subtypes over a mean follow-up of 6 years [38] or 9.6 years [39].

ASSOCIATIONS BETWEEN PUFAs PLASMA / TISSUE LEVELS AND COGNITIVE DYSFUNCTION OR DEMENTIA

Measurements of n-3 PUFAs in plasma or on erythrocyte membrane are represented as good markers of their status in the body, which represent as short-term (30 days) or long-term (60-90 days) dietary intake respectively and are generally considered as better estimates than intake estimated from dietary questionnaires [40-42]. Several case-control studies have sought to investigate associations between these levels and dementia or mild cognitive impairment (Table **2**). Compared with controls, plasma levels of C20:5n-3, DHA and total n-3 fatty acids, and n-3/n-6 ratio have been found to be lower in people with AD, other dementia, and MCI, and total n-6 fatty acids levels have been found to be higher in AD and MCI groups only [43-45].

Table 2: Summarising reports of association between PUFAs plasma/tissue levels and cognitive dysfunction or dementia.

	Study design	Sample	Number	Fish/or PUFAs measurement	Cognitive domains or other measurements	Principal findings
Conquer *et al.* 2000 [43]	Case-control study	AD (mean age 83y), other dementia (79y), CIND (83y), and normal controls (76y).	84(19 AD, 10 other dementia, 36 CIND, 19 normal controls	Plasma fatty acids composition of various PL fractions (total PL, PC, PE, lysoPC)	Dementia and other dementia (DSM-IV), AD(NINDCS-ADRDA), CIND (DSM-IV and neuro-psychological exam)	Lowered plasma PL and PC levels of C20:5n-3, DHA and total n-3 fatty acids, and n-3/n-6 ratio in the AD, other dementia, and CIND group. Higher total n-6 fatty acids levels in AD and CIND groups only
Tully *et al.* 2000 [44]	Case-control study	AD (mean age 76.5y) and controls (70y)	Case: 148 Control: 45	Serum cholesteryl ester-DHA level	AD(NINDCS-ADRDA)	Lowered serum EPA and DHA in AD group. DHA and total saturated fatty acids levels is a determinant to MMSE and CDR.
Whalley *et al.* 2004 [46]	Nested case-control study	Age 64y, subgroup from a UK cohort with childhood IQ test.	120 (60 fish oil users, 60 nonusers)	Fatty acids composition on erythrocyte membrane	A single composite score of neuropsychological tests	Total n-3 PUFAs, EPA, DHA, and n3/n6 correlate with IQ at childhood and at 64 y. DHA/AA correlates to IQ at 64y after childhood IQ controlled.
Cherubini *et al.* 2007 [45]	Case-control study, community study	Mean age 75y	935 (57 dementia, 153CIND, 725 normal cognition)	Plasma fatty acids	Dementia(DSM-III-R), CIND(MMSE<23 and some disability related to cognitive impairment)	Lowered plasma n-3 PUFAs and ALA levels in dementia group.
Heude *et al.* 2003 (EVA study, France) [47]	Prospective design with 4-y follow-up	Age 63-74 y	246	Fatty acids composition of erythrocyte membrane	Moderate cognitive decline(decrease of MMSE$\geqq$2 over a 4-y period	Higher n-3 PUFAs, higher DHA but not EPA, and lower n-6 PUFAs associates with a lower risk of cognitive decline.
Beydoun *et al.* 2007 (Atherosclerosis Risk in Community Study, USA) [49]	Prospective design, 1987-1998	Age$\geqq$50 y (mean 56-57 y)	2,251	Plasma fatty acids in cholesteryl esters and phospholipids	Decline in 3 separate cognitive tests and a composite of them	Higher risk in global cognitive decline associates with high palmitic acid and high AA, and low LA. High n-3 PUFAs reduces risk of decline in verbal fluency, particularly in hypertensive and dyslipidemic subjects.
Whalley *et al.* 2008 (UK) [48]	Prospective design with 4-y follow-up	Mean age 64.4(0.6)y, subgroup from a UK cohort with IQ test at 11 and 64y;	120	Fatty acids content of erythrocyte membrane	A principal component from 5 cognitive test	Cognitive benefits associates with higher erythrocyte n-3 PUFAs only in APOE ε 4 non-carriers. The benefits of DHA disappear after adjustment.

Table 2: cont....

Lauri *et al.* 2003 (Canadian Study of Health and Aging, Canada) [52]	Cross-section and prospective study with 5-y follow-up	Baseline participants with 52 dementia, 43 CIND, and 79 normal cognition, mean age of normal cognition group: 76.9 y	174 at baseline, 79 normal cognition group for cohort	Plasma concentration of fatty acids	PL Dementia(DSM-III-R), AD(NINDCS-ADRDA), Vascular dementia(ICD-10), CIND(modified version of Zaudig's criteria)	No significant difference in n-3 PUFAs levels between groups at baseline. In prospective analysis, higher EPA in CIND group and higher n-3 PUFAs and higher total PUFAs levels in dementia cases compared to controls.
Schaefer *et al.* 2006 (Framingham Heart Study, USA) [50]	Prospective design with 9.1- y follow-up	Age 55-88y (mean 76y), no dementia.	899	Plasma fatty acids	Dementia(DSM-IV) and AD(NINDCS-ADRDA), CDR≥1	The top quartile of plasma PC DHA level associates with a significant 47% reduction in the risk of developing all-cause dementia. No other significant association.
Samieri *et al.* 2008 (Three City study, France) [51]	Prospective design with 4-y follow-up	Age ≥ 65y, no dementia	1,214	Plasma fatty acids	Dementia (DSM-IV)	Higher EPA associates with a lower incidence of dementia, independently of depressive status. Higher ratio of AA/DHA and n-6/n-3 fatty acids relates to an increased risk of dementia, particularly in depressive subjects.
Kroger *et al.* 2009 (Canadian Study of Health and Aging, Canada) [53]	Prospective design with 4.9-y follow-up	Age ≥ 65y, no dementia	663	Fatty acids profiles on erythrocyte membrane, blood mercury concentration	Dementia(DSM-IV), AD(NINDCS-ADRDA)	No associations between n-3 PUFAs, EPA, or DHA and dementia or AD.

CIND, cognitively impaired nondemented; DSM, Diagnostic and Statistical Manual of Mental Disorders; ICD-10, World Health Organization International Classfication of Diseases; NINDCS –ADDTC, National Institute of Neurological and Communication Disorders and Stroke-Association International pour la Recherche et l' Enseignement en Neurosciences; NINDCS- ADRDA, National Institute of Neurological and Communication Disorders and Stroke- Alzheimer's Disease and Related Disorder; PL, phospholipids; PC, phosphotidylcholine; PE, phosphotidylethanolamine.

These findings suggest that associations with decreased levels of plasma DHA are not specific to AD, but also to other dementia and MCI. In addition, DHA and total saturated fatty acids levels correlate well with cognitive and functional scores [44]. A nested case control study have also found that total erythrocyte n-3 PUFAs and DHA/AA were positively associated with cognitive function in late life after adjusting for childhood IQ [46]. In prospective studies, higher proportions of total n-3 PUFAs on the erythrocyte membrane were found to be associated with a lower risk of cognitive decline [47;48], while higher proportions of saturated and total n-6 fatty acids were associated with greater risk of cognitive decline [47]. One study found that higher cognitive function associated with higher erythrocyte n-3 PUFAs was only significant in people without the APOE ε 4 allele. In addition, these associations with cognition have been found to be stronger for n-3 PUFAs rather than DHA [48].

In the Atherosclerosis Risk In Communities study, plasma n-3 PUFAs levels have been revealed no association with global cognitive decline, but is associated with a lower risk of decline in verbal fluency, particularly among hypertensive and dyslipidemic subjects and in those who are less depressed at baseline [49].

Among prospective studies of dementia incidence, the top quartile of plasma phosphatidylcholine DHA level was associated with a reduction in the risk of developing all-cause dementia in Framingham Heart Study. This association appeared to be stronger for all-cause dementia than for AD after considering possible confounders [50]. In the French Three-City Study, higher plasma EPA but not DHA levels were associated with a lower incidence of dementia, independent of depressive symptoms at baseline; in addition, a higher ratio of AA/DHA and n-6/n-3 fatty acids were associated with an increased risk of dementia, particularly in people with depressive symptoms at baseline [51].

In contrast to other cross-section studies, there was no significant difference in n-3 PUFAs plasma concentrations between controls, MCI, and dementia in one study [52] and, in the prospective analysis, higher rather than lower EPA levels were found among MCI incident cases compared to controls as well as higher n-3 and total PUFAs levels among subjects who later developed dementia [52]. No associations between n-3 PUFAs, EPA, or DHA levels and dementia or AD were found in the Canadian Study of Health and Aging [53].

ANIMAL MODELS AND BIOLOGICAL STUDIES

Numerous studies have sought to investigate the different possible mechanisms underlying associations between PUFAs levels and cognitive function in animal models, particularly in those of AD. DHA administration has been found to improve impairment of cognitive performance in young and aged amyloid β-infused rats [54;55]. Significantly increased DHA content and DHA/AA ratio in both the hippocampus and the cerebral cortex have been found in these, and the ratio has been found to be negatively correlated with learning ability as well [54]. DHA-induced alterations in synaptic plasma membrane fluidity have been suggested to contribute to improvement of cognitive function, supported by findings from one study that the annular to non-annular fluidity ratio of the synaptic plasma membrane was positively correlated with total avoidance learning. [56]. DHA supplementation have been suggested to decrease Aβ deposition in AD animal models [57-60]. In addition to DHA, Ethyl-EPA [61-63] has been reported to improve memory impairment in rat models, though these effects have been suggested due to EPA as a DHA precursor [63]. AA has been reported to correlate with long-term potentiation (LTP) in memory process [64], and AA administration to aged animals has been suggested to alleviate age-related cognitive deficits in rats [65].

Some other possible biological effects of PUFAs have also been reported. DHA supplementation may slightly increase relative cerebral blood flow [57], and decrease the number of activated hippocampal microglia in APP/PS1 transgenic mice [66]. Furthermore, dietary DHA administration has been found to be associated with increased cerebral acetylcholine levels and improved cognitive performance in stroke-prone spontaneously hypertensive rats [67], further supported by findings that a chronic ALA deficient diet was associated with reduced release of acetylcholine by KCL-stimulation, mimicking neuron activation, in hippocampus and reduced muscarinic receptor binding in rats [68]. DHA has also been reported to suppress increases in cortical and hippocampal lipid peroxide levels and reactive oxygen species in Aβ infused rats, suggesting an antioxidative action [69].

Importantly, a novel DHA-derived 10,17 S-docosatriene, neuroprotectin D1 (NPD1), has been found recently [70]. NPD1 may repress Aβ42-triggered activation of pro-inflammatory genes, while upregulating anti-apoptotic genes and soluble amyloid precursor protein-α stimulated NPD1 biosynthesis from DHA. These findings suggest that NPD1 may promote brain cell survival via the induction of anti-apoptotic and neuroprotective gene expression programmes to suppress Aβ42-induced neurotoxicity [71]

In contrast, in a study in AD transgenic and non-transgenic mice, n-3 supplementation was not associated with significant cognitive benefit to either group and did not affect brain soluble or insoluble Aβ levels. In this study, brain levels of n-6 rather than n-3 fatty acids were strongly correlated with cognitive impairment for both mouse groups [72]. The authors further hypothesized that use of fish oil supplements would not protect against AD, at least in high-risk individuals, but that normal individuals might benefit from high omega-3 intake if it corrects an elevation in the brain levels of n-6 fatty acids as a result [72].

CLINICAL TRIALS OF FISH OIL/OR PUFAs SUPPLEMENTATION – EFFECTS ON COGNITIVE FUNCTION OR DEMENTIA PROGRESSION

Despite the evidence from observational research of possible protective effects of fish oil/n-3 PUFAs on cognitive impairment and/or risk of dementia, there have been few clinical trials published to date (Table **3**). An early open-label uncontrolled trial showed that EPA 900mg/day administration in the patients with AD improved the scores of Mini Mental State Examination (MMSE) with maximal effects at 3 months, sustained benefits over 6 months, but decreased effects after 6 months [73]. There were no clinically important treatment effects of ethyl-EPA on cognition in patients with AD during a 12-week treatment period in a pilot open-label study [74]. However, an open-labeled study in people with mild to moderate dementia attributed to thrombotic cerebrovascular disease found that cognitive function had improved in the intervention group after 3 and 6 months of DHA supplementation compared to normal diet comparison [75]. Another supplementation study administered 240mg AA and DHA in participants with MCI, organic brain lesions, and AD compared with olive oil in participants with MCI for 90 days. Only patients with MCI and organic brain lesion in the AA+DHA group showed cognitive improvement while compared to their baseline performance [76].

Table 3: Summarising clinical trials reporting effects of polyunsaturated fatty acids on cognitive function.

	Study design and duration	Sample	Number	Supplementation	Cognitive domains or other measurements	Principal findings
Terano *et al.* 1999 [75]	Open-label trial for 1y; control group	Elderly with mild to moderate thrombotic cerebrovascular dementia	20	Ten with DHA 720mg/day; ten without DHA (usual diet)	MMSE HDS-R	Improvement of scores of MMSE and HDS-R in DHA group after 3 and 6 months supplementation. Unchanged or lowered MMSE in control group.
Otsuka *et al.* 2000 [73]	Open-label trial for 4 to 12 months (average 8.3±5.8 months)	Patients with AD	27	Ethyl EPA 900mg/day	MMSE	Improvement of MMSE scores with maximal effects at 3 months and lasted 6 months. Diminished effects after 6 months.
Boston *et al.* 2004 [74]	Open-label trial: 12 weeks without treatment followed by 12-week treatment	Patients with mild to moderate AD	22	Ethyl EPA 900mg/day	MMSE ADAS-Cog Visual analogue rating scales for functioning	Little difference between treatment and baseline period in cognitive decline. Only small improvement in carer's visual analogue rating.
Kotani *et al.* 2006 [76]	Open-label, placebo-controlled trial for 90 days	21 patients with MCI, 10 patients with organic brain lesions, and 8 patients with AD	39	MCI-P(n=9) with olive oil 240mg/day; MCI-A(n=12) and others with AA + DHA 240mg/day	Japanese version of RBANS: 12 subtest with 5 domains of cognitive function	Improvement of immediate memory and attention in MCI-A group, and immediate and delayed memories in organic group. No significant improvement of any score in AD and MCI-P groups.
Freund-Levi *et al.* 2006 [78]	Randomized, double-blind placebo-controlled for 6 months, followed fish oil for at least 6 months	Participants with mild to moderate AD	174	DHA 1720mg and EPA600mg /day; Placebo: 1000mg corn oil(including 600mg LA)	MMSE ADAS-cog	No benefits of n3 supplementation for reduction of cognitive decline in this group. Positive effects in a subgroup (n=32) with very mild cognitive dysfunction (MMSE>27 points).
Chiu *et al.* 2008 [77]	Randomized, double-blind placebo-controlled for 24 weeks	23 participants with MCI and 23 with mild to moderate AD	46	EPA 1080 mg and DHA 720 mg/day; placebo: olive oil	ADAS-cog CIBIC-plus	No significant difference in ADAS-cog change. Treatment group with better improvement on CIBIC-plus scale. N3 fatty acids group with significant improvement in ADAS-cog in patient with MCI but not dementia.
Van de Rest *et al.* 2008 [80]	Randomized, double-blind placebo-controlled study for 26 weeks	Individuals aged 65 years or older with relatively preserved cognitive function (MMSE≧21)	302	High dose EPA+DHA 1,800mg/day; Low-dose EPA+DHA 400mg/d; or placebo(high-oleic sunflower oil)	Extensive cognitive test battery with five tests	No significant effects on any cognitive domain for low- or high-dose fish oil group. A possible effects of n3 PUFAs on attention in APOE ε 4 carriers and in male subjects in secondary analysis.

ADAS-cog, cognitive portion of Alzheimer's Disease Assessment Scale; CIBIC-plus, Clinician's Interview-Based Impression of Change scale; HDRS, Hasegawa's Dementia rating scale; RBANS, repeatable battery for the assessment of neuropsychological functions.

A double-blind randomized placebo controlled trial in patients with mixed of AD or MCI found the n-3 PUFAs intervention group to have better improvement in global clinical function compared to placebo over a 24-week follow-up in spite of no significant group difference in cognitive function change. In a secondary sub-group analysis, the intervention group showed significantly greater improvement in cognitive function compared to the placebo group in participants with MCI, which was not observed in those with AD. In addition, higher proportions of EPA on RBC membranes were associated with better cognitive outcome [77]. A larger randomized placebo-controlled trial investigated effects on cognition and behavior of combination of DHA and EPA in 174 people with mild to moderate AD over a 6 month follow-up period; in this study, decline in cognitive function did not differ in the two groups [78]. However, in a subgroup with very mild cognitive dysfunction (MMSE >27), a significant reduction in MMSE decline rate was found in the intervention compared to placebo group. Supplementation did not have marked effects on activities of daily living, caregiver burden, or neuropsychiatric symptoms except a beneficial effect in a sub-group analysis on depressive symptoms in APOE ε 4 non-carriers and agitation symptoms in APOE ε 4 carriers [79]. In another, double-blind randomized placebo-controlled trial the effects of high-dose and low-dose EPA+DHA, and placebo were compared in 300 individuals aged 65 years or older with relatively preserved cognitive function over a 26-week period [80]. In spite of good compliance, there were no significant group differences in any of the cognitive domains. In a secondary sub-group analysis, APOE ε 4 carriers in the fish oil group and male participants in the low-dose group showed significant improvement in attentional function compared with placebo group.

Possible Mechanisms in Fish Oil/n-3 PUFAs in Cognition

Some possible mechanisms to explain the associations of PUFAs with cognition have been hypothesized although remain controversial.

First, DHA supplementation may decrease Aβ deposition, supported by animal models as previously described [3;57-60]. Since the production and accumulation of Aβ peptide is believed to be central to the pathogenesis of AD, any decrease its deposition should further reduce consequences, such as oxidation and lipid peroxidation, glutamatergic excitoxicity, inflammation, and activation of biochemical cascades underlying apoptotic cell death [6].

Second, anti-inflammatory and anti-oxidant effects have been suggested [81] which may also reduce the impact of AD pathology. Chronic inflammatory processes has been implicated in the neuropathology of AD, although these may differ at different stages of AD, including MCI [82]. AA, the principal precursor for eicosanoid synthesis, can form prostaglandins, thromboxanes, leukotrienes and further regulate the production of other mediators including inflammatory cytokines [83]. The anti-inflammatory actions of n-3 PUFAs may result from decreasing the amount of AA in cell membranes (and so available for eicosanoid production), from decreasing the production of the classic inflammatory cytokines, and from decreasing the expression of adhesion molecules involved in inflammatory interactions between leukocytes and endothelial cells [83]. Oxidative damage is also elevated in patients with AD, and n-3 PUFAs, as highly unsaturated fatty acids, are vulnerable to oxidative attack via autocatalytic lipid peroxidation reactions. However, DHA may promote antioxidant defenses by potentiation of Akt, and n-3 PUFAs have been found to reduce oxidative reaction in AD animal models and humans [84;85].

Third, the negative association between n-3 PUFAs and cognitive decline or dementia may result from beneficial effects on cardiovascular disease. Accumulating evidence has suggested that vascular risk factors contribute to AD, and it appears that there is a spectrum of dementia outcomes associated with cerebrovascular disease ranging from pure vascular dementia syndromes to pure AD but including a large majority of patients with contributions from both Alzheimer and vascular pathologies [86]. N-3 PUFAs are being actively evaluated in the primary and secondary prevention of cardiovascular diseases, especially cardiac infarction and heart failure [87]. Reduction in risk of thrombosis or stroke may be important in suspected protective effects on risk of cognitive decline or dementia. Relatively low levels of plasma DHA have been found to be associated not only with AD, but also with other dementia [43] – i.e. an apparently non-specific association. DHA supplementation specifically to patients with mild to moderate dementia from thrombotic cerebrovascular disease has also been suggested to be beneficial [75].

Fourth, AA and DHA could enhance acetylcholine release, which may modulate LTP and synaptic plasticity and further improve cognition. Although there are limitations in the cholinergic hypothesis for cognitive dysfunction in AD [88], cholinesterase inhibitors are still the most prescribed agents for patients with mild to moderate AD [6] and cholinergic neurotransmission is recognised to be critical for processes underlying arousal, attention, learning and memory [89]. A DHA-enriched phospholipid diet has been found to enhance the spontaneous and evoked release of acetylcholine by the hippocampus of aging rats [90], and hippocampus acetylcholine levels have been found to be correlated with cognitive performance [67].

Finally, other possible mechanisms have also been hypothesized [81;91]. It has been suggested that n-3 PUFAs may potentiate NMDA responses, then trigger LTP, and memory formation and consolidation [91], supported by an observed decrease in NMDA receptors after supplementation of an n-3 depleted diet in AD transgenic mouse [92]. Direct effects of NPD1 may regulate the brain cell survival and repair involving neurotrophic, anti-apoptotic and anti-inflammatory signaling. It may also have important regulatory interactions with the molecular genetic mechanisms affecting βamyloid precursor protein and amyloid peptide neurobiology [71].

In short, the evidence accumulated to date cannot be treated separately since the biological and neurochemical substrates for memory and disorders underlying dementia are highly complex and far from being understood. Neurobiological explanations for n-3 PUFAs in cognitive protection, if true, are unlikely to involve a single mechanism and may well depend on between-mechanism interactions, such as NPD1 explaining the possible actions of DHA via anti-inflammatory and anti-oxidant effects on amyloid deposition and further apoptotic cascades.

Some Considerations Regarding the Role of n-3 PUFAs in Cognition

Considerations in Epidemiological Studies and Tissue Composition Studies

Although most epidemiological studies and tissue compositions studies seem to support a positive association between fish / or n-3 PUFAs intake and decreased risk of dementia or cognitive decline, methodological limitations make the inconsistent results difficult to interpret. These include a range of different exposures and outcomes being evaluated. Some researchers have used fish consumption as their exposure, while others have used n-3 PUFAs intake or n-3 PUFAs levels in plasma or on erythrocyte membranes. In terms of cognitive outcomes, some have used a diagnostic system to define dementia or AD, whereas others have created their own definitions of cognitive impairment or decline. It is uncertain whether these different measurements represent the same exposure or outcome. For example, people with the same fish consumption may not get the same n-3 PUFAs levels because of individual variability in metabolism influenced in turn by genes, age and consumption of other diets [93].

As well as exposure and outcome definitions, the time for follow-up has differed considerably, which will affect the incidence rate of dementia as well as selective attrition. This might account for different results in three papers analyzing the same database over different time periods [33;38;39]. Positive findings in short-term prospective studies may well be due to unrecognized dementia at baseline and reflect reverse causation where people with preclinical disease have a poorer diet with lower fish intake [39].

Some confounders are also difficult to exclude. For example, the validity of different diet questionnaires and whether a baseline diet represents long-term habit are argued though interviewer-administered frequent food questionnaire has been proved to be a validated tool used in patients with MCI and AD [94]. Also, people who eat more fish may have other healthier lifestyles such as higher consumption of vegetables, fruits, health supplements, increased exercise, and other aspects of a healthy lifestyle. All of these may influence the cognitive function but are difficult to control for in many observation studies. Finally, the way fish is cooked or the type of fish intake may also influence the results [36] but have not been registered in most studies.

Considerations in Clinical Trials

The composition and dosage of PUFAs used in the clinical trials have differed across previous studies. Some trials have supplemented with combination of EPA and DHA, others with EPA or AA only, and others with combination of DHA and AA. In addition, the PUFAs dosage used in trials has varied from 240 to 2300mg/day. Some observational studies have suggested there is a dose-response trend [26;30], but others have not found this. Whether there is a therapeutic window is unknown. Although there is a stronger link between DHA and cognition function epidemiologically and biologically, the results are not so consistent for other PUFAs, EPA and AA, also found to be associated with cognition.

Consistent with an observational finding of higher plasma EPA but not DHA concentration associated with a lower incidence of dementia [51], our supplementation trial also found an association between EPA rather than DHA levels on erythrocyte membranes and cognitive function [77]. Some animal models also suggest that EPA may have a protective function in cognition [61-63]. N-6 PUFAs, on the other hand, have been suggested to be hazardous to health and regular consumption of n-6 rich oils not compensated by n-3 rich oils was found to be associated with an increased risk of dementia among APOEε4 non-carriers [37]. However, AA may have a role in LTP and AA administration to aged animals was also found to alleviate age-related deficits in spatial cognition [65]. A more balancing ratio of ALA / LA has been suggested to be most effective in improving learning in animal models [95]. In order to achieve more consistency between trials, further investigation is required into the efficacy of EPA or AA as add-ons to DHA as well as to establish the adequate dosage of these PUFAs.

Gene-Environment Interactions – APOEε4

In the Cardiovascular Health Cognition Study and the Three-City cohort study, consumption of fatty fish was found to be associated with a reduced risk of dementia and AD only in APOEε4 non-carriers [36;37]. This was further supported by the finding that cognitive benefits in higher erythrocyte n-3 PUFAs content were only significant in the APOEε4 non-carriers [48]. Nevertheless, others have not been able to find this interaction and contradictory results have also been reported, such as a study which found that only APOEε4 carriers had an improvement in attention after fish oil supplementation[80] and another study finding that intake of ALA was associated with reduced risk of AD in APOEε4 carriers but not non-carriers [35]. Further research is clearly required to establish such interactions, not only with APOE genotype but for other candidates as well.

Potential Candidate Samples for n-3 PUFAs supplementation Trials

In spite of inconsistent results in clinical trials, some important issues have been raised by research to date. First, n-3 PUFAs may have more demonstrable effects on cognition in people with mild cognitive impairment [76;77] or with dementia but very mild cognitive dysfunction [78]. In addition, there was no significant cognitive effects of fish oil administration in older people with relatively preserved cognitive function [80], although the 26 week follow-up period may not have been sufficient to detect effects.

The association between n-3 fatty acids and cognitive decline has been suggested to be strongest in middle-aged people with hypertension and hyperlipidemia [32] which might reflect benefits on cardiovascular status and possibly a rationale for targeting groups at high risk of stroke and vascular dementia. Previously, it has been suggested that the potential antithrombotic effect of n-3 PUFAs may increase the risk of bleeding; however, according to Harris's systematic review of 19 clinical trials with n-3 PUFAs supplementation for patients with a high risk of bleeding, the risk of clinically significant bleeding was stated to be "virtually nonexistent" [96]. People with depression, known to be at higher risk of developing dementia, may also be a target sample, since a higher ratio of AA/DHA and n-6/n-3 PUFAs were associated with an increased risk of dementia, particularly in depressive subjects [51].

Taken together, the most suitable candidates for further clinical trials of n-3 PUFAs supplementation may be samples who are particularly vulnerable to cognitive decline. People with MCI or mild dementia, with vascular disease/risk, with late-life depression, are therefore possible candidates.

Combinations with Other Nutritional Supplements?

In addition to fish consumption, other components of the diet such as fruits, vegetables, and wine, which are rich in antioxidant nutrients, or other nutrient supplements, such as S-adenosylmethionine, vitamin B, calcium and others, have found to be associated with lower risk of dementia [37;97;98]. Previously, some studies have suggested that the association with cognition is stronger for fish consumption than n-3 PUFAs, which suggested that there may be other micronutrients other than n-3 PUFAs which could benefit cognitive function. In the Three-City study, a combination of dietary sources of omega-3 PUFAs and antioxidants seemed to be more effective for modifying dementia risk [37]. Whether other positive nutritional factors, such as antioxidants, are also needed to combine with n-3 PUFAs to exert more influence on cognition needs further investigation.

CONCLUSIONS

Emerging data from epidemiological research, tissue composition studies, animal models, and clinical trials has yet to provide a strong enough evidence to support n-3 PUFAs supplementation in patients with cognitive impairment or dementia. Contradictory findings in observation studies may reflect level of adjustment for possible confounders, inconsistencies in measurements in exposure and outcome, and in follow-up duration. Further biological research is required to clarify potential underlying mechanisms of action, while larger and better designed trials are needed for evaluating interventions. As well as larger sample size and better design, trials should consider evaluating mixed compositions and dosage of PUFAs, including ratio of n3/n6, mixture of EPA and DHA, and monotherapy of DHA.

REFERENCES

[1]   Ferri, C. P.; Prince, M.; Brayne, C.; Brodaty, H.; Fratiglioni, L.; Ganguli, M.; Hall, K.; Hasegawa, K.; Hendrie, H.; Huang, Y.; Jorm, A.; Mathers, C.; Menezes, P. R.; Rimmer, E. Scazufca. Global prevalence of dementia: a Delphi consensus study. *Lancet*, **2005**, *366*, 2112-2117.

[2]   Prince, M.; Graham, N.; Brodaty, H.; Rimmer, E.; Varghese, M.; Chiu, H.; Acosta, D.; Scazufca, M. Alzheimer Disease

International's 10/66 Dementia Research Group - one model for action research in developing countries. *Int. J. Geriatr. Psychiatry*, **2004**, *19*, 178-181,

[3]   Rabins, P. V.; Blacker, D.; Rovner, B. W.; Rummans, T.; Schneider, L. S.; Tariot, P. N.; Blass, D. M.; McIntyre, J. S.; Charles, S. C.; Anzia, D. J.; Cook, I. A.; Finnerty, M. T.; Johnson, B. R.; Nininger, J. E.; Schneidman, B.; Summergrad, P.; Woods, S. M.; Berger, J.; Cross, C. D.; Brandt, H. A.; Margolis, P. M.; Shemo, J. P.; Blinder, B. J.; Duncan, D. L.; Barnovitz, M. A.; Carino, A. J.; Freyberg, Z. Z.; Gray, S. H.; Tonnu, T.; Kunkle, R.; Albert, A. B.; Craig, T. J.; Regier, D. A.; Fochtmann, L. J. American Psychiatric Association practice guideline for the treatment of patients with Alzheimer's disease and other dementias. Second edition. *Am. J. Psychiatry*, **2007**, *164*, 5-56.

[4]   Burdge, G. C.; Calder, P. C. Conversion of alpha-linolenic acid to longer-chain polyunsaturated fatty acids in human adults. *Reprod. Nutr. Dev.*, **2005**, *45*, 581-597.

[5]   Shahidi, F.; Miraliakbari, H. Omega-3 (n-3) fatty acids in health and disease: Part 1-cardiovascular disease and cancer. *J. Med. Food*, **2004**, *7*, 387-401.

[6]   Cummings, J. L. Alzheimer's disease. *N. Engl. J. Med.*, **2004**, *351*, 56-67.

[7]   Hebert, L. E.; Scherr, P. A.; Bienias, J. L.; Bennett, D. A.; Evans, D. A. Alzheimer disease in the US population: prevalence estimates using the 2000 census. *Arch. Neurol.*, **2003**, *60*, 1119-1122.

[8]   Gandy, S. The role of cerebral amyloid beta accumulation in common forms of Alzheimer disease. *J. Clin. Invest.*, **2005**, *115*, 1121-1129.

[9]   Myers, R. H.; Schaefer, E. J.; Wilson, P. W.; D'Agostino, R.; Ordovas, J. M.; Espino, A.; Au, R.; White, R. F.; Knoefel, J. E.; Cobb, J. L.; McNulty, K. A.; Beiser, A.; Wolf, P. A. Apolipoprotein E epsilon4 association with dementia in a population-based study: The Framingham study. *Neurology*, **1996**, *46*, 673-677.

[10]   Bennett, D. A.; Schneider, J. A.; Wilson, R. S.; Bienias, J. L.; Arnold, S. E. Neurofibrillary tangles mediate the association of amyloid load with clinical Alzheimer disease and level of cognitive function. *Arch. Neurol.*, **2004**, *61*, 378-384.

[11]   Geldmacher, D. S. Treatment guidelines for Alzheimer's disease: redefining perceptions in primary care. *Prim. Care. Companion J. Clin. Psychiatry*, **2007**, *9*, 113-121.

[12]   Gauthier, S.; Reisberg, B.; Zaudig M.; Petersen, R. C.; Ritchie, K.; Broich, K.; Belleville, S.; Brodaty, H.; Bennett, D.; Chertkow, H.; Cummings, J L.; de Leon, M.; Feldman, H.; Ganguli, M.; Hampel, H.; Scheltens, P.; Tierney, M. C.; Whitehouse, P.; Winblad, B. Mild cognitive impairment. *Lancet*, **2006**, *367*, 1262-1270.

[13]   McNamara, R. K.; Carlson, S. E. Role of omega-3 fatty acids in brain development and function: potential implications for the pathogenesis and prevention of psychopathology. *Prostaglandins Leukot. Essent. Fatty Acids*, **2006**, *75*, 329-349.

[14]   Clandinin, M. T.; Chappell, J. E.; Leong, S.; Heim, T.; Swyer, P. R.; Chance, G.W. Intrauterine fatty acid accretion rates in human brain: implications for fatty acid requirements. *Early. Hum. Dev.*, **1980**, *4*, 121-129.

[15]   Carver, J. D.; Benford, V. J.; Han, B.; Cantor, A. B. The relationship between age and the fatty acid composition of cerebral cortex and erythrocytes in human subjects. *Brain. Res. Bull.*, **2001**, *56*, 79-85.

[16]   Hoffman, D. R.; Boettcher, J. A.; Diersen-Schade, D. A. Toward optimizing vision and cognition in term infants by dietary docosahexaenoic and arachidonic acid supplementation: a review of randomized controlled trials. *Prostaglandins Leukot. Essent. Fatty Acids*, **2009**, *81*, 151-158.

[17]   Simmer, K.; Patole, S. K.; Rao, S. C. Long chain polyunsaturated fatty acid supplementation in infants born at term. *Cochrane. Database. Syst. Rev.*, **2008**, CD000376.

[18]   Yehuda, S.; Rabinovitz, S.; Carasso, R. L.; Mostofsky, D. I. The role of polyunsaturated fatty acids in restoring the aging neuronal membrane. *Neurobiol. Aging*, **2002**, *23*, 843-853.

[19]   Favrelere, S.; Stadelmann-Ingrand, S.; Huguet, F.; De Javel, D.; Piriou, A.; Tallineau, C.; Durand, G. Age-related changes in ethanolamine glycerophospholipid fatty acid levels in rat frontal cortex and hippocampus. *Neurobiol. Aging*, **2000**, *21*, 653-660.

[20]   Ulmann, L.; Mimouni, V.; Roux, S.; Porsolt, R.; Poisson, J. P. Brain and hippocampus fatty acid composition in phospholipid classes of aged-relative cognitive deficit rats. *Prostaglandins Leukot. Essent. Fatty Acids*, **2001**, *64*, 189-195.

[21]   Yehuda, S.; Rabinovitz, S.; Carasso, R. L.; Mostofsky, D. I. Fatty acid mixture counters stress changes in cortisol, cholesterol, and impair learning. *Int. J. Neurosci.*, **2000**, *101*, 73-87.

[22]   Borsonelo, E. C.; Galduroz, J. C. The role of polyunsaturated fatty acids (PUFAs) in development, aging and substance abuse disorders: review and propositions. *Prostaglandins Leukot. Essent. Fatty Acids*, **2008**, *78*, 237-245.

[23]   Terracina, L.; Brunetti, M.; Avellini, L.; De Medio, G. E.; Trovarelli, G.; Gaiti, A. Arachidonic and palmitic acid utilization in aged rat brain areas. *Mol. Cell. Biochem.*, **1992**, *115*, 35-42.

[24]   Kalmijn, S.; van Boxtel, M. P.; Ocke, M.; Verschuren, W. M.; Kromhout, D.; Launer, L. J. Dietary intake of fatty acids and fish in relation to cognitive performance at middle age. *Neurology*, **2004**, *62*, 275-280.

[25]   Ortega, R. M.; Requejo, A. M.; Andres, P.; Lopez-Sobaler, A. M.; Quintas, M. E.; Redondo, M. R.; Navia, B.; Rivas, T.

Dietary intake and cognitive function in a group of elderly people. *Am. J. Clin. Nutr.*, **1997**, *66*, 803-809.

[26] Nurk, E.; Drevon, C. A.; Refsum, H.; Solvoll, K.; Vollset, S. E.; Nygard, O.; Nygaard, H. A.; Engedal, K.; Tell, G. S.; Smith, A. D. Cognitive performance among the elderly and dietary fish intake: the Hordaland Health Study. *Am. J. Clin. Nutr.*, **2007**, *86,* 1470-1478.

[27] Albanese, E.; Dangour, A. D.; Uauy, R.; Acosta, D.; Guerra, M.; Guerra, S. S.; Huang, Y.; Jacob, K. S.; de Rodriguez, J. L.; Noriega, L. H.; Salas, A.; Sosa, A. L.; Sousa, R. M.; Williams, J.; Ferri, C. P.; Prince, M. J. Dietary fish and meat intake and dementia in Latin America, China, and India: a 10/66 Dementia Research Group population-based study. *Am. J. Clin. Nutr.*, **2009**, *90*, 392-400.

[28] Kalmijn, S.; Feskens, E. J.; Launer, L. J.; Kromhout, D. Polyunsaturated fatty acids, antioxidants, and cognitive function in very old men. *Am. J. Epidemiol*, **1997**, *145*, 33-41.

[29] Morris, M. C.; Evans, D. A.; Tangney, C. C.; Bienias, J. L.; Wilson, R. S. Fish consumption and cognitive decline with age in a large community study. *Arch. Neurol.*, **2005**, *62*, 1849-1853.

[30] van Gelder, B. M.; Tijhuis, M.; Kalmijn, S.; Kromhout, D. Fish consumption, n-3 fatty acids, and subsequent 5-y cognitive decline in elderly men: the Zutphen Elderly Study. *Am. J. Clin. Nutr.*, **2007**, *85*, 1142-1147.

[31] Vercambre, M. N.; Boutron-Ruault, M. C.; Ritchie, K.; Clavel-Chapelon, F.; Berr, C. Long-term association of food and nutrient intakes with cognitive and functional decline: a 13-year follow-up study of elderly French women. *Br. J. Nutr.*, **2009**, *102*, 419-427.

[32] Beydoun, M. A.; Kaufman, J. S.; Sloane, P. D.; Heiss, G.; Ibrahim, J. n-3 Fatty acids, hypertension and risk of cognitive decline among older adults in the Atherosclerosis Risk in Communities (ARIC) study. *Public Health Nutr.*, **2008**, *11*, 17-29.

[33] Kalmijn, S.; Launer, L. J.; Ott, A.; Witteman, J. C.; Hofman, A.; Breteler, M. M. Dietary fat intake and the risk of incident dementia in the Rotterdam Study. *Ann. Neurol.*, **1997**, *42*, 776-782.

[34] Barberger-Gateau, P.; Letenneur, L.; Deschamps, V.; Peres, K.; Dartigues, J. F.; Renaud, S. Fish, meat, and risk of dementia: cohort study. *BMJ*, **2002**, *325*, 932-933.

[35] Morris, M. C.; Evans, D. A.; Bienias, J. L.; Tangney, C. C.; Bennett, D. A.; Wilson, R.S.; Aggarwal, N.; Schneider, J. Consumption of fish and n-3 fatty acids and risk of incident Alzheimer disease. *Arch. Neurol.*, **2003**, *60*, 940-946.

[36] Huang, T L.; Zandi, P. P.; Tucker, K. L.; Fitzpatrick, A. L.; Kuller, L. H.; Fried, L. P.; Burke, G. L.; Carlson, M. C. Benefits of fatty fish on dementia risk are stronger for those without APOE epsilon4. *Neurology*, **2005**, *65*, 1409-1414.

[37] Barberger-Gateau, P.; Raffaitin, C.; Letenneur, L.; Berr, C.; Tzourio, C.; Dartigues, J. F.; Alperovitch, A. Dietary patterns and risk of dementia: the Three-City cohort study. *Neurology*, **2007**, *69*, 1921-1930.

[38] Engelhart, M. J.; Geerlings, M. I.; Ruitenberg, A.; Van Swieten, J. C.; Hofman, A.; Witteman, J. C.; Breteler, M. M. Diet and risk of dementia: Does fat matter?: The Rotterdam Study. *Neurology*, **2002**, *59*, 1915-1921.

[39] Devore, E. E.; Grodstein, F.; van Rooij, F. J.; Hofman, A.; Rosner, B.; Stampfer, M. J.; Witteman, J. C.; Breteler, M. M. Dietary intake of fish and omega-3 fatty acids in relation to long-term dementia risk. *Am. J. Clin. Nutr.*, **2009**, *90*, 170-176.

[40] Bjerve, K. S.; Brubakk, A. M.; Fougner, K. J.; Johnsen, H.; Midthjell, K.; Vik, T. Omega-3 fatty acids: essential fatty acids with important biological effects, and serum phospholipid fatty acids as markers of dietary omega 3-fatty acid intake. *Am. J. Clin. Nutr.*, **1993**, *57*, 801S-805S.

[41] Ma, J.; Folsom, A. R.; Shahar, E.; Eckfeldt, J. H. Plasma fatty acid composition as an indicator of habitual dietary fat intake in middle-aged adults. The Atherosclerosis Risk in Communities (ARIC) Study Investigators. *Am. J. Clin. Nutr.*, **1995**, *62*, 564-571.

[42] Katan, M. B.; Deslypere, J. P.; van Birgelen, A. P.; Penders, M.; Zegwaard, M. Kinetics of the incorporation of dietary fatty acids into serum cholesteryl esters, erythrocyte membranes, and adipose tissue: an 18-month controlled study. *J. Lipid Res.*, **1997**, *38*, 2012-2022.

[43] Conquer, J. A.; Tierney, M. C.; Zecevic, J.; Bettger, W. J.; Fisher, R. H. Fatty acid analysis of blood plasma of patients with Alzheimer's disease, other types of dementia, and cognitive impairment. *Lipids*, **2000**, *35*, 1305-1312.

[44] Tully, A. M.; Roche, H. M.; Doyle, R.; Fallon, C.; Bruce, I.; Lawlor, B.; Coakley, D.; Gibney, M. J. Low serum cholesteryl ester-docosahexaenoic acid levels in Alzheimer's disease: a case-control study. *Br. J. Nutr.*, **2003**, *89*, 483-489.

[45] Cherubini, A.; Andres-Lacueva, C.; Martin, A.; Lauretani, F.; Iorio, A. D.; Bartali, B.; Corsi, A.; Bandinelli, S.; Mattson, M.P.; Ferrucci, L. Low plasma N-3 fatty acids and dementia in older persons: the InCHIANTI study. *J. Gerontol. A. Biol. Sci. Med. Sci.*, **2007**, *62*, 1120-1126.

[46] Whalley, L. J.; Fox, H. C.; Wahle, K. W.; Starr, J. M.; Deary, I. J. Cognitive aging, childhood intelligence, and the use of food supplements: possible involvement of n-3 fatty acids. *Am. J. Clin. Nutr.*, **2004**, *80*, 1650-1657.

[47] Heude, B.; Ducimetiere, P.; Berr, C. Cognitive decline and fatty acid composition of erythrocyte membranes--The EVA Study. *Am. J. Clin. Nutr.*, **2003**, *77,* 803-808.

[48]   Whalley, L. J.; Deary, I. J.; Starr, J. M.; Wahle, K. W.; Rance, K. A.; Bourne, V. J.; Fox, H.C. n-3 Fatty acid erythrocyte membrane content, APOE varepsilon4, and cognitive variation: an observational follow-up study in late adulthood. *Am. J. Clin. Nutr.*, **2008**, *87*, 449-454.

[49]   Beydoun, M. A.; Kaufman, J. S.; Satia, J. A.; Rosamond, W.; Folsom, A. R. Plasma n-3 fatty acids and the risk of cognitive decline in older adults: the Atherosclerosis Risk in Communities Study. *Am. J. Clin. Nutr.*, **2007**, *85*, 1103-1111.

[50]   Schaefer, E. J.; Bongard, V.; Beiser, A. S.; Lamon-Fava, S.; Robins, S. J.; Au, R.; Tucker, K. L.; Kyle, D. J.; Wilson, P. W.; Wolf, P. A. Plasma phosphatidylcholine docosahexaenoic acid content and risk of dementia and Alzheimer disease: the Framingham Heart Study. *Arch. Neurol.*, **2006**, *63*, 1545-1550.

[51]   Samieri, C.; Feart, C.; Letenneur, L.; Dartigues, J. F.; Peres, K.; Auriacombe, S.; Peuchant, E.; Delcourt, C.; Barberger-Gateau, P. Low plasma eicosapentaenoic acid and depressive symptomatology are independent predictors of dementia risk. *Am. J. Clin. Nutr.*, **2008**, *88*, 714-721.

[52]   Laurin, D.; Verreault, R.; Lindsay, J.; Dewailly, E.; Holub, B. J. Omega-3 fatty acids and risk of cognitive impairment and dementia. *J. Alzheimers Dis.*, **2003**, *5*, 315-322.

[53]   Kroger, E.; Verreault, R.; Carmichael, P. H.; Lindsay, J.; Julien, P.; Dewailly,E.; Ayotte, P.; Laurin, D. Omega-3 fatty acids and risk of dementia: the Canadian Study of Health and Aging. *Am.J. Clin. Nutr.*, **2009**, *90*, 184-192.

[54]   Gamoh, S.; Hashimoto, M.; Hossain, S.; Masumura, S. Chronic administration of docosahexaenoic acid improves the performance of radial arm maze task in aged rats. *Clin. Exp. Pharmacol Physiol.*, **2001**, *28*, 266-270.

[55]   Hashimoto, M.; Tanabe, Y.; Fujii, Y.; Kikuta, T.; Shibata, H.; Shido, O. Chronic administration of docosahexaenoic acid ameliorates the impairment of spatial cognition learning ability in amyloid beta-infused rats. *J. Nutr.*, **2005**, *135*, 549-555.

[56]   Hashimoto, M.; Hossain, S.; Shimada, T.; Shido, O. Docosahexaenoic acid-induced protective effect against impaired learning in amyloid beta-infused rats is associated with increased synaptosomal membrane fluidity. *Clin. Exp. Pharmacol. Physiol.*, **2006**, *33*, 934-939.

[57]   Hooijmans, C. R.; Van der Zee, C. E.; Dederen, P. J.; Brouwer, K. M.; Reijmer, Y. D.; van Groen, T.; Broersen, L. M.; Lutjohann, D.; Heerschap, A.; Kiliaan, A. J. DHA and cholesterol containing diets influence Alzheimer-like pathology, cognition and cerebral vasculature in APPswe/PS1dE9 mice. *Neurobiol. Dis.*, **2009**, *33*, 482-498.

[58]   Oksman, M.; Iivonen, H.; Hogyes, E.; Amtul, Z.; Penke, B.; Leenders, I.; Broersen, L.; Lutjohann, D.; Hartmann, T.; Tanila, H. Impact of different saturated fatty acid, polyunsaturated fatty acid and cholesterol containing diets on beta-amyloid accumulation in APP/PS1 transgenic mice. *Neurobiol. Dis.*, **2006**, *23*, 563-572.

[59]   Green, K. N.; Martinez-Coria, H.; Khashwji, H.; Hall, E. B.; Yurko-Mauro, K. A.; Ellis, L.; LaFerla, F. M. Dietary docosahexaenoic acid and docosapentaenoic acid ameliorate amyloid-beta and tau pathology via a mechanism involving presenilin 1 levels. *J. Neurosci.*, **2007**, *27*, 4385-4395.

[60]   Hashimoto, M.; Hossain, S.; Agdul, H.; Shido, O. Docosahexaenoic acid-induced amelioration on impairment of memory learning in amyloid beta-infused rats relates to the decreases of amyloid beta and cholesterol levels in detergent-insoluble membrane fractions. *Biochim. Biophys. Acta.*, **2005**, *1738*, 91-98.

[61]   Song, C.; Phillips, A. G.; Leonard, B. E.; Horrobin, D. F. Ethyl-eicosapentaenoic acid ingestion prevents corticosterone-mediated memory impairment induced by central administration of interleukin-1beta in rats. *Mol. Psychiatry*, **2004**, *9*, 630-638.

[62]   Song, C.; Horrobin, D. Omega-3 fatty acid ethyl-eicosapentaenoate, but not soybean oil, attenuates memory impairment induced by central IL-1beta administration. *J. Lipid. Res*, **2004**, *45*, 1112-1121.

[63]   Hashimoto, M.; Hossain, S.; Tanabe, Y.; Kawashima, A.; Harada, T.; Yano, T.; Mizuguchi, K.; Shido, O. The protective effect of dietary eicosapentaenoic acid against impairment of spatial cognition learning ability in rats infused with amyloid beta ((1-40)). *J. Nutr. Biochem.*, **2008**.

[64]   Lynch, M. A.; Voss, K. L. Membrane arachidonic acid concentration correlates with age and induction of long-term potentiation in the dentate gyrus in the rat. *Eur. J. Neurosci.*, **1994**, *6*, 1008-1014.

[65]   Okaichi, Y.; Ishikura, Y.; Akimoto, K.; Kawashima, H.; Toyoda-Ono, Y.; Kiso, Y.; Okaichi, H. Arachidonic acid improves aged rats' spatial cognition. *Physiol. Behav.*, **2005**, *84*, 617-623.

[66]   Oksman, M.; Iivonen, H.; Hogyes, E.; Amtul, Z.; Penke, B.; Leenders, I.; Broersen, L.; Lutjohann, D.; Hartmann, T.; Tanila, H. Impact of different saturated fatty acid, polyunsaturated fatty acid and cholesterol containing diets on beta-amyloid accumulation in APP/PS1 transgenic mice. *Neurobiol. Dis.*, **2006**, *23*, 563-572.

[67]   Minami, M.; Kimura, S.; Endo, T.; Hamaue, N.; Hirafuji, M.; Togashi, H.; Matsumoto, M.; Yoshioka, M.; Saito, H.; Watanabe, S.; Kobayashi, T.; Okuyama, H. Dietary docosahexaenoic acid increases cerebral acetylcholine levels and improves passive avoidance performance in stroke-prone spontaneously hypertensive rats. *Pharmacol. Biochem. Behav.*, **1997**, *58*, 1123-1129.

[68]   Aid, S.; Vancassel, S.; Poumes-Ballihaut, C.; Chalon, S.; Guesnet, P.; Lavialle, M. Effect of a diet-induced n-3 PUFA

depletion on cholinergic parameters in the rat hippocampus. *J. Lipid Res.*, **2003**, *44*, 1545-1551.

[69] Hashimoto, M.; Hossain, S.; Shimada, T.; Sugioka, K.; Yamasaki, H.; Fujii, Y.; Ishibashi, Y.; Oka, J.; Shido, O. Docosahexaenoic acid provides protection from impairment of learning ability in Alzheimer's disease model rats. *J. Neurochem.*, **2002**, *81*, 1084-1091.

[70] Mukherjee, P. K.; Marcheselli, V. L.; Serhan, C. N.; Bazan, N. G. Neuroprotectin D1: a docosahexaenoic acid-derived docosatriene protects human retinal pigment epithelial cells from oxidative stress. *Proc. Natl. Acad. Sci. U. S. A.*, **2004**, *101*, 8491-8496.

[71] Lukiw, W. J.; Cui, J. G.; Marcheselli, V. L.; Bodker, M.; Botkjaer, A.; Gotlinger, K.; Serhan, C. N.; Bazan, N. G. A role for docosahexaenoic acid-derived neuroprotectin D1 in neural cell survival and Alzheimer disease. *J. Clin. Invest.*, **2005**, *115*, 2774-2783.

[72] Arendash, G. W.; Jensen, M. T.; Salem, N. Jr.; Hussein, N.; Cracchiolo, J.; Dickson, A.; Leighty, R.; Potter, H. A diet high in omega-3 fatty acids does not improve or protect cognitive performance in Alzheimer's transgenic mice. *Neuroscience*, **2007**, *149*, 286-302.

[73] Otsuka, M. Analysis of dietary factors in Alzheimer's disease: clinical use of nutritional intervention for prevention and treatment of dementia. *Nippon. Ronen. Igakkai. Zasshi.*, **2000**, *37*, 970-973.

[74] Boston, P. F.; Bennett, A.; Horrobin, D. F.; Bennett, C. N. Ethyl-EPA in Alzheimer's disease--a pilot study. *Prostaglandins Leukot. Essent. Fatty Acids*, **2004**, *71*, 341-346.

[75] Terano, T.; Fujishiro, S.; Ban, T.; Yamamoto, K.; Tanaka, T.; Noguchi, Y.; Tamura, Y.; Yazawa, K.; Hirayama, T. Docosahexaenoic acid supplementation improves the moderately severe dementia from thrombotic cerebrovascular diseases. *Lipids*, **1999**, *34*, S345-S346.

[76] Kotani, S.; Sakaguchi, E.; Warashina, S.; Matsukawa, N.; Ishikura, Y.; Kiso, Y.; Sakakibara, M.; Yoshimoto, T.; Guo, J.; Yamashima, T. Dietary supplementation of arachidonic and docosahexaenoic acids improves cognitive dysfunction. *Neurosci. Res.*, **2006**, *56*, 159-164.

[77] Chiu, C. C.; Su, K. P.; Cheng, T. C.; Liu, H. C.; Chang, C. J.; Dewey, M. E.; Stewart, R.; Huang, S. Y. The effects of omega-3 fatty acids monotherapy in Alzheimer's disease and mild cognitive impairment: a preliminary randomized double-blind placebo-controlled study. *Prog. Neuropsychopharmacol. Biol. Psychiatry*, **2008**, *32*, 1538-1544.

[78] Freund-Levi, Y.; Eriksdotter-Jonhagen, M.; Cederholm, T.; Basun, H.; Faxen-Irving, G.; Garlind, A.; Vedin, I.; Vessby, B.; Wahlund, L. O.; Palmblad, J. Omega-3 fatty acid treatment in 174 patients with mild to moderate Alzheimer disease: OmegAD study: a randomized double-blind trial. *Arch. Neurol.*, **2006**. *63*. 1402-1408.

[79] Freund-Levi, Y.; Basun, H.; Cederholm, T.; Faxen-Irving, G.; Garlind, A.; Grut, M.; Vedin, I.; Palmblad, J.; Wahlund, L. O.; Eriksdotter-Jonhagen, M. Omega-3 supplementation in mild to moderate Alzheimer's disease: effects on neuropsychiatric symptoms. *Int. J. Geriatr. Psychiatry*, **2008** *23*, 161-169.

[80] van de Rest, O.; Geleijnse, J. M.; Kok, F. J.; van Staveren, W. A.; Dullemeijer, C.; Olderikkert, M.G.; Beekman, A. T.; de Groot, C. P. Effect of fish oil on cognitive performance in older subjects: a randomized, controlled trial. *Neurology*, **2008**, *71*, 430-438.

[81] Cole, G. M.; Ma, Q.L.; Frautschy, S. A. Omega-3 fatty acids and dementia. *Prostaglandins Leukot. Essent. Fatty Acids*, **2009**, *81*, 213-221.

[82] Magaki, S.; Mueller, C.; Dickson, C.; Kirsch, W. Increased production of inflammatory cytokines in mild cognitive impairment. *Exp. Gerontol.*, **2007**, *42*, 233-240.

[83] Calder, P. C. Polyunsaturated fatty acids and inflammation. *Prostaglandins Leukot. Essent. Fatty Acid*, **2006**, *75*, 197-202.

[84] Calon, F.; Lim, G. P.; Yang, F.; Morihara, T.; Teter, B.; Ubeda, O.; Rostaing, P.; Triller, A.; Salem, N. Jr.; Ashe, K. H.; Frautschy, S. A.; Cole, G. M. Docosahexaenoic acid protects from dendritic pathology in an Alzheimer's disease mouse model. *Neuron*, **2004**, *43*, 633-645.

[85] Mori, T. A.; Puddey, I. B.; Burke, V.; Croft, K. D.; Dunstan, D. W.; Rivera, J. H.; Beilin, L. J. Effect of omega 3 fatty acids on oxidative stress in humans: GC-MS measurement of urinary F2-isoprostane excretion. *Redox. Rep.*, **2000**, *5*, 45-46.

[86] Viswanathan, A.; Rocca, W. A.; Tzourio, C. Vascular risk factors and dementia: how to move forward?.*Neurology*, **2009**, *72*, 368-374.

[87] Lavie, C. J.; Milani, R. V.; Mehra, M. R.; Ventura, H. O. Omega-3 polyunsaturated fatty acids and cardiovascular diseases. *J. Am. Coll. Cardiol.*, **2009**, *54*, 585-594.

[88] Robbins, T. W.; McAlonan, G.; Muir, J. L.; Everitt, B. J. Cognitive enhancers in theory and practice: studies of the cholinergic hypothesis of cognitive deficits in Alzheimer's disease. *Behav. Brain Res.*, **1997**, *83*, 15-23.

[89] Blokland, A. Acetylcholine: a neurotransmitter for learning and memory? *Brain Res. Brain Res. Rev.*, **1995**, *21*, 285-300.

[90] Favreliere, S.; Perault, M. C.; Huguet, F.; De Javel, D.; Bertrand, N.; Piriou, A.; Durand, G. DHA-enriched phospholipid diets modulate age-related alterations in rat hippocampus. *Neurobiol. Aging*, **2003**, *24*, 233-243.

[91]   Das, U. N. Folic acid and polyunsaturated fatty acids improve cognitive function and prevent depression, dementia, and Alzheimer's disease--but how and why? *Prostaglandins Leukot. Essent. Fatty Acids*, **2008**, *78*, 11-19.

[92]   Calon, F.; Lim, G. P.; Morihara, T.; Yang, F.; Ubeda, O.; Salem, N. Jr.; Frautschy, S. A.; Cole, G. M. Dietary n-3 polyunsaturated fatty acid depletion activates caspases and decreases NMDA receptors in the brain of a transgenic mouse model of Alzheimer's disease. *Eur. J. Neurosci.*, **2005**, *22*, 617-626.

[93]   Fotuhi, M.; Mohassel, P.; Yaffe, K. Fish consumption, long-chain omega-3 fatty acids and risk of cognitive decline or Alzheimer disease: a complex association. *Nat. Clin. Pract. Neurol.*, **2009**, *5*, 140-152.

[94]   Arsenault, L. N.; Matthan, N.; Scott, T. M.; Dallal, G.; Lichtenstein, A. H.; Folstein, M. F.; Rosenberg, I.; Tucker, K. L. Validity of estimated dietary eicosapentaenoic acid and docosahexaenoic acid intakes determined by interviewer-administered food frequency questionnaire among older adults with mild-to-moderate cognitive impairment or dementia. *Am. J. Epidemiol.*, **2009**, *170*, 95-103.

[95]   Yehuda, S.; Carasso, R. L. Modulation of learning, pain thresholds, and thermoregulation in the rat by preparations of free purified alpha-linolenic and linoleic acids: determination of the optimal omega 3-to-omega 6 ratio. *Proc. Natl. Acad. Sci. U. S.A.*, **1993**, *90*, 10345-10349.

[96]   Harris, W. S. Expert opinion: omega-3 fatty acids and bleeding-cause for concern? *Am. J. Cardiol.*, **2007**, *99*, 44C-46C.

[97]   Panza, F.; Frisardi, V.; Capurso, C.; D'Introno, A.; Colacicco, A. M.; Di Palo, A.; Imbimbo, B. P.; Vendemiale, G.; Capurso, A.; Solfrizzi, V. Polyunsaturated fatty acid and S-adenosylmethionine supplementation in predementia syndromes and Alzheimer's disease: a review. *Scientific World Journal*, **2009**, *9*, 373-389.

[98]   Velho, S.; Marques-Vidal, P.; Baptista, F.; Camilo, M. E. Dietary intake adequacy and cognitive function in free-living active elderly: a cross-sectional and short-term prospective study. *Clin. Nutr.*, **2008**, *27*, 77-86.

CHAPTER 4

Beneficial Roles of the n-3 Long-Chain Polyunsaturated Fatty Acids on the Management of Obesity and Metabolic Syndrome

Mardia López-Alarcón, M.D., Ph.D.[1],*, Mariela Bernabé-García, M.Sc.[2], and Javier Mancilla-Ramírez, M.D., Ph.D.

[1]*Unit of Research in Medical Nutrition, Pediatric Hospital, "Centro Médico Nacional Siglo XXI", Mexican Institute of Social Security, Mexico City, México and* [2]*Instituto Nacional de Perinatología "Isidro Espinoza de los Reyes", Mexico City, Mexico*

Abstract: n-3 Long-chain polyunsaturated fatty acids (n-3 LC-PUFAs) reduce inflammation through several mechanisms; therefore, components of the metabolic syndrome (MetS) may be treated with n-3 LC-PUFAs not only because of their anti-inflammatory effect but also through other mechanisms that reduce vascular reactivity, oxidative stress, and insulin resistance. In this chapter, the positive effects of n-3 LC-PUFAs on each component of the MetS are analyzed. Although information regarding the effect of n-3 LC-PUFAs on obesity is discordant, it seems to either inhibit weight gain or decrease inflammatory status; thus, a positive effect is expected. Regarding the effect on hypertension, evidence shows that treatment with n-3 LC-PUFAs reduces blood pressure by modulating vascular reactivity and reducing arterial thickness. In the case of dyslipidemia, the triacylglycerol-lowering properties of n-3 LC-PUFAs are among the best established *in vivo* actions. These fatty acids also decrease the assembly and secretion of very-low-density lipoproteins (VLDL), increase the conversion of VLDL to LDL particles, and increase β-oxidation of other fatty acids in mitochondria. Finally, insulin sensitivity is improved by treatment with n-3 LC-PUFAs due to actions on lipid dysregulation, adiponectin production and by their role as ligands to peroxisome proliferator-activated receptors. In conclusion, evidence mainly from randomized trials supports the positive effect of n-3 LC-PUFAs in all four components of the MetS. Such beneficial roles act at the cellular and the molecular levels and have been demonstrated in animal and human experimental studies and at the community level. Therefore, use of n-3 LC-PUFAs may be widely recommended to decrease the MetS and subsequently decrease the risk for type 2 diabetes and cardiovascular disease.

1. INTRODUCTION

Metabolic syndrome (MetS) is a cluster of metabolic alterations that in combination predicts the development of cardiovascular disease (CVD) and type 2 diabetes mellitus (T2DM) [1]. According to the Third National Cholesterol Education Program Adult Treatment Panel (NCEP-ATP III), components of MetS are obesity or central adiposity, hypertension, dyslipidemia, and insulin resistance, considering this last component as the underlying cause of the syndrome [2]. Using these criteria, the prevalence of MetS in adults in the U.S. has been estimated at 23.7% [3]. In children, a similar pattern of increase has been observed. Recent studies report that the prevalence of MetS in children and adolescents has increased significantly [4–6], which is relevant, because higher increases for the adult population and secondary increases in the risk for CVD and T2DM are expected. In the elderly, the prevalence of MetS is even higher, varying from 11 to 55% depending on the criteria used for diagnosis, i.e., NCEP or WHO. Interestingly, in the elderly population, obesity and hypertension are the most prevalent individual components of MetS [7].

As previously mentioned, markers of metabolic distress that comprise MetS are central obesity, hypertension, dyslipidemia, and insulin resistance. These metabolic alterations are all related to inflammation and oxidative stress [8]. Therefore, in theory, their management may be approached with anti-inflammatory and antioxidant agents.

N-3 Long-chain polyunsaturated fatty acids (n-3 LC-PUFAs), particularly eicosapentaenoic acid (EPA) and docosahexaenoic acid (DHA), have been shown to reduce inflammation and oxidative status through several

*Address correspondence to Dra. Mardia López-Alarcón: Unidad de Investigación Médica en Nutrición, Apartado Postal C-029 "Coahuila", Coahuila #5, Col. Roma, México, D.F., 06703, México. Teléfono/Fax (52) 55 5627 6944; E-mail: marsau2@prodigy.net.mx

mechanisms, i.e., reducing expression of the proinflammatory cytokines IL-1, IL-6 and TNF among others, decreasing the adhesion and migration of monocytes, promoting interactions between leukocytes and endothelial cells, and modulating the activation of transcription factors such as NF-κB and peroxisome proliferator-activated receptor (PPARs) [9–13]. Thus, based on the knowledge about the underlying inflammatory status of patients with MetS it is expected that treatment with n-3 LC-PUFAs would decrease MetS manifestations. This is of high priority because, by decreasing the expression of MetS components, a secondary decrease in the risk for CVD and T2DM may be achieved.

BENEFICIAL EFFECTS OF N-3 LC-PUFAs ON THE COMPONENTS OF MetS

Obesity and n-3 LC-PUFAs

Adipose tissue (AT) is an active endocrinological tissue that secretes multiple metabolically important proteins known as adipokines [14]. At the cellular level, obesity is not solely a disturbance of AT because there are other cell types that participate as well. Such is the case of infiltrating macrophages in AT that makes obesity comparable to a low-grade chronic inflammation with links between adipose cells and cells from the immune system. Thus, AT in obese individuals produce increased amounts of the inflammatory cytokines IL-1, IL-6, TNF, and leptin, either from AT alone or from the infiltrated macrophages. However, in any case, these cytokines are responsible for the inflammatory status of obese individuals and, at least in part, for the adverse effects of obesity [15,16].

The information regarding the effect of n-3 LC-PUFAs supplementation on obesity is discordant. Multiple studies have demonstrated that the concentration of inflammatory molecules in obese individuals decreases after treatment with n-3 LC-PUFAs [17,18], but a direct role on weight loss is controversial. Experimental studies have reported weight loss in animals exposed to diets with fish oil, but epidemiological studies in humans have failed to demonstrate such a positive effect. For obesity, as for the other components of MetS, the positive effects of n-3 LC-PUFAs supplementation seem to depend on the form in which the fatty acids are administered, i.e., fish oil or pure preparations of ethyl esters [19], as well as the doses used [20]. Interestingly, it seems that in experimental animals the n-3 LC-PUFAs supplementation reduces body fat in already obese animals that were exposed to obesogenic diets [21]. Similar supplementation prevents weight gain when humans are exposed to high-fat diets instead of enhanced weight loss after dieting. This effect is exerted through several mechanisms, e.g., by inhibiting the key enzymes responsible for lipid synthesis including fatty acid synthase and stearoyl-CoA desaturase-1, preventing free fatty acids from entering adipocytes for lipogenesis, and through suppressive effects on key factors involved in adipocyte differentiation and fat storage [22,23].

In any case, if treatment with n-3 LC-PUFAs decreases body weight, reduces the inflammatory status, or prevents weight gain, the net result is a positive effect on the risk of obesity and potentially reduces the risk for T2DM and CVD.

Hypertension and n-3 LC-PUFAs

The finding that n-3 LC-PUFAs can produce reductions in blood pressure in experimental models of hypertension and in humans, although moderate, has focused attention on a potential modulating role of these fatty acids on vascular contraction and dilatation, together with an effect on cell proliferation [11,24,25].

Endothelium plays a key role in vascular function and maintains homeostasis via the production of a range of biochemical mediators. Endothelial cells generate vasoactive agents such as angiotensin II, nitric oxide (NO), endothelial-derived hyperpolarizing factor (EDHF), eicosanoids, and endothelin (ET), as well as producing increased levels of free radicals. Thus, endothelial dysfunction potentially reflects an imbalance between vasoconstrictor factors (angiotensin II, endothelin, eicosanoids (isoprostane, TxA_2, and PGH_2 and the superoxide anion O_2), and vasodilator compounds (NO, EDHF, and eicosanoid PGF_2) [24]. Moreover, increased thickness of the arterial wall, which is characteristic of hypertension, is caused by abnormal growth and hypertrophy of vascular smooth muscle cells [26]. Therefore, those factors that reduce vessel reactivity and vascular wall thickness may be beneficial.

N-3 LC-PUFAs have the ability to modify eicosanoid synthesis toward a more vasodilatation profile and to increase endogenous NO production [25,26]. Both processes can influence vascular reactivity and improve arterial elasticity. In a meta-analysis that included only controlled trials, treatment with n-3 LC-PUFAs supplementation lowered systolic and diastolic blood pressure in hypertensive patients [27,28] and in prehypertensive rats [29]. In another study, 6-week DHA treatment also attenuated the development of hypertension in prehypertensive rats. Similarly, 6-week treatment with DHA reduced vascular wall thickness in the coronary artery and aorta of hypertensive rats [26]. In humans, a meta-analysis of 30 randomized trials found that fish oil intake reduced heart rate, which is one of the factors that produce an increase of wall thickness by 1.6 beats/min [30].

Thus, it seems that treatment with n-3 LC-PUFAs modulates vessel reactivity and inhibits growth and proliferation of smooth muscle cells. In addition to the effects on vascular reactivity and arterial thickness, other mechanisms may contribute to the beneficial effects of n-3 LC-PUFAs supplementation. For instance, such supplementation reduces the synthesis of aldosterone, modulates calcium release from vascular smooth muscle cells, and activates ATP-sensitive potassium channels [29]. Interestingly, at the molecular level, n-3 LC-PUFAs have the ability to adjust the expression of pro-atherogenic genes such as endothelial leukocyte adhesion molecules and decrease the expression of pro-inflammatory genes including cyclooxygenase-2 and inflammatory cytokines [11,24,29].

In summary, evidence that supplementation with n-3 LC-PUFAs reduces blood pressure is highly robust. This evidence comes from animal studies and from studies at the community level. Moreover, the reported studies demonstrate that such fatty acids act at different levels in the physiopathology of systemic arterial hypertension. Therefore, recommendation for n-3 LC-PUFAs supplementation to hypertensive patients is a focus of current research.

Dyslipidemia and n-3 LC-PUFAs

Clinical data of dyslipidemia that are considered for diagnosis of MetS are high levels of plasma triglycerides (TG) and low HDL-cholesterol [31]. Other alterations in lipid metabolism comprise high total cholesterol and LDL-cholesterol but are not included as components of MetS.

Several studies show the potent TG-lowering effect of n-3 LC-PUFAs in both normo- and hyperlipidemic subjects [32–36]. Results from two meta-analyses report that treatment with combinations of EPA and DHA decreases by 25-34% the plasma concentration of TG [32,33]. More recent studies demonstrated even higher decreases (30–50%) in TG levels [34–36]. In contrast, little or no effect on HDL and LDL cholesterol has been reported. Actually, an increase in LDL cholesterol was reported by other authors [37,38].

TG-lowering properties are among the best established *in vivo* actions of n-3 LC-PUFAs. In addition, other important positive effects have been identified on lipid metabolism. For instance, n-3 LC-PUFAs also decrease the assembly and secretion of VLDL, increase the conversion of VLDL to LDL particles, increase β-oxidation of other fatty acids in mitochondria and peroxisomes possibly through activation of PPAR-α, and stimulate lipoprotein lipase to remove TG-rich lipoproteins via changes in gene expression in adipose tissue [23,25,39].

It has been reported that high apolipoprotein (apo) B is associated with atherogenesis, whereas apo A-1 is present in atheroprotective particles. Therefore, some authors also recommend monitoring the concentration of these apoproteins and their ratio in patients with MetS. Limited information exists in relation to the effect of n-3 LC-PUFAs supplementation on apoA- and apoB-containing lipoproteins. In a recent study conducted in diabetic patients, it was reported that a combination of monounsaturated fatty acids (MUFAs) with EPA and DHA resulted in a lower postprandial increase of apoB- and apoC-containing lipoproteins as compared to the increase in subjects who received a diet with MUFA plus α-linolenic acid, suggesting an additional positive effect of n-3 LC-PUFAs supplementation on these lipoprotein particles [40].

Thus, based on the TG-reducing effect of n-3 LC-PUFAs, the U.S. Food and Drug Administration has approved a prescription form of n-3 LC-PUFAs, which is administered as an adjunct to an appropriate diet for the treatment of very high TG levels [25]. However, according to current evidence, it seems adequate to prescribe such fatty acids also for slightly and moderately elevated TG levels.

Insulin Resistance and n-3 LC-PUFAs

Insulin resistance is a decreased response of glucose to the action of insulin in skeletal muscle and adipose tissue. In the first stage, the combination of inhibited glucose uptake and decreased insulin action leads to increased insulin secretion. In more advanced stages, these factors lead to alterations in glucose homeostasis and T2DM [1]. This definition takes into account the effects of insulin on glucose metabolism, but in the last decade a more "lipocentric" point of view has been considered. According to this "lipocentric" hypothesis, abnormalities in fatty acids metabolism may result in an inappropriate accumulation of lipids in muscle, liver, and β-cells, which produces a "lipotoxic" status. This ectopic lipid accumulation, mainly that occurring in the intramyocellular space, eventually generates insulin resistance [15,41].

Thus, obesity is strongly related to decreased insulin sensitivity due to the ectopic fatty acids metabolites such as long-chain acyl-CoA and diacylglycerol, which accumulate within myocytes [41], as well as to the inflammatory status of obese individuals [42]. The inflammatory response has been linked to the development of insulin resistance in two different ways. First, activation of inflammatory signaling intermediates may be directly involved in serine phosphorylation of the insulin receptor substrate-1 (IRS-1) within insulin-sensitive cells, leading to alterations in insulin action and, second, because of the inflammatory cell infiltration within adipose tissue. Such infiltration is involved in altering adipocyte lipid metabolism, for instance, TNF-α is reported to promote lipolysis [15,16].

Other relevant links between obesity and insulin resistance have been mentioned. For instance, white adipose tissue expansion of obese individuals occurs mainly through saturated fatty acid accumulation, which also causes metabolic dysregulation [43]. Excess of the saturated fatty acid palmitate not only expands white adipose tissue but also increases inflammation through endoplasmic reticulum stress and generation of reactive oxygen species. In fact, *in vivo* and *in vitro* studies have shown that palmitate inhibits the activation of IRS-1 and phosphatidyl-inositol-3-kinase, producing insulin resistance [44]. In addition to the metabolic dysregulation, obese subjects also express low levels of adiponectin, which is an insulin-sensitizing protein produced by adipocytes. Saturated fatty acids seem to impair insulin sensitivity by reducing adiponectin secretion and impairing insulin signaling pathways required for glucose uptake [43,45].

Finally, peroxisome proliferator-activated receptors (PPARs), which are members of a nuclear receptor superfamily that participates in glucose and lipid metabolism, are also implicated as targets for n-3 fatty acids. The three isoforms of PPARs have different tissue distributions, physiological roles, and ligand specificities; i.e., activation of PPARα lowers lipid levels, PPARγ increases insulin sensitivity, and PPARδ regulates cholesterol and glucose levels [46]. Recent evidence suggests that n-3 LC-PUFAs protect against insulin resistance through activation of the PPARα isotype, which first would decrease blood lipid concentrations and subsequently would decrease intracellular lipid abundance [47]. PPARα and PPARγ are the natural ligands for EPA and DHA within physiological concentrations. Therefore, their activation acts in combination to improve insulin sensitivity.

Together this information suggests that by altering the fatty acid profile in muscle and AT, i.e., replacing saturated fatty acids by polyunsaturated fatty acids, specifically the n-3 long-chain type, it is likely to improve insulin sensitivity because these will reduce inflammation, enhance insulin signaling pathways, and increase adiponectin levels.

Example of the Effect of DHA and EPA Administration to Prepubertal and Pubertal Children on Inflammatory Status and Insulin Resistance

In a study conducted by our research group with the aim of elucidating whether insulin resistance is an inflammatory condition that results from obesity, we followed a sample of 76 obese children with fasting insulin ≥15 μg/mL and HOMA ≥6.5. Children were randomly assigned to receive a daily dose of 350 mg DHA plus 560 mg EPA (Omega III, Salmon Oil, GNLD) or placebo for 1 month. At study initiation, children were asked not to alter their eating patterns, in order to maintain body weight. Fasting serum glucose, insulin, IL-6, TNFα, TNF-soluble receptors (sTNFR1 and sTNFR2) and C-reactive protein concentrations were determined at inclusion and after a 1-month follow-up. Despite the fact that no changes in body weight or body composition were observed in any group, only children who received n-3 LC-PUFAs decreased their insulin concentration by 14% and HOMA value by 20% (Figs. **1** and **2**). Similarly, some indicators of inflammation, including IL-6 and TNF-soluble receptor, were decreased in the group of children who received n-3 LC-PUFAs (Fig. **2**). In contrast, no changes were observed in the group of children receiving placebo.

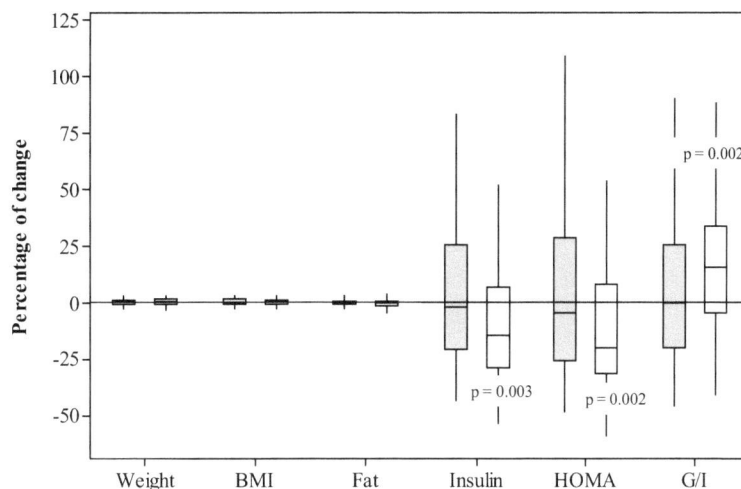

Figure 1: Despite no changes in body weight and body composition, children who received n-3 LC-PUFAs ☐ presented decrease in all the indicators of insulin resistance such as fasting insulin concentration (p = 0.003), the HOMA value (p = 0.002), and the glucose:insulin ratio (p = 0.002. The group of children who received placebo ☐ did not show any change.

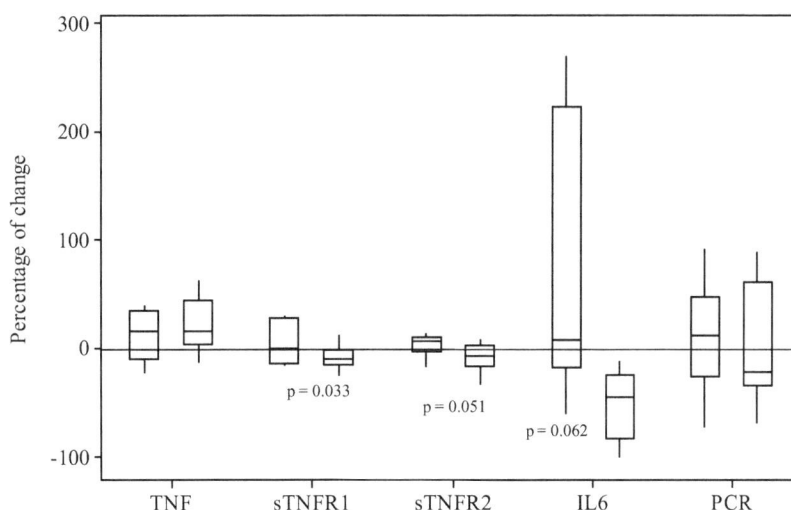

Figure 2: After 1-month supplementation, children who received the n-3 LC-PUFAs ☐ demonstrated a significant decrease in some indicators of inflammation such as IL-6 and the soluble receptors of TNF, sTNRF1 and sTNRF2. No changes were observed in TNFa concentration or C-reactive protein. No changes were present either in any cytokines in the group of children who received placebo ☐.

Evidence from this study demonstrated that n-3 LC-PUFAs supplementation decreased inflammation and improved insulin sensitivity. Because no changes in body weight or body composition were observed, this study also helps to discriminate between the effects of weight reduction from the effect of n-3 LC-PUFAs supplementation on insulin resistance.

In conclusion, evidence mainly from randomized trials supports the positive effect of n-3 LC-PUFAs supplementation on all four components of MetS. Such beneficial roles act at the cellular and molecular levels and have been demonstrated in animal and human experimental studies, as well as at the community level. Therefore, n-3 LC-PUFAs supplementation may be widely recommended to decrease the clinical manifestations of MetS and, secondarily, risk for T2DM and CVD.

REFERENCES

[1]   DeFronzo, R. A. Insulin resistance: a multifaceted syndrome responsible for NIDDM, obesity, hypertension, dyslipidaemia and atherosclerosis. *Neth. J. Med.*, **1997**, *50*, 191-197.

[2]   Expert Panel on Detection, Evaluation and Treatment of High Blood Cholesterol in Adults. Executive summary of the Third Report of the National Cholesterol Education Program (NCEP) Expert Panel on Detection, Evaluation and Treatment of High Blood Cholesterol in Adults (Adult Treatment Panel III). *JAMA*, **2001**, *285*, 2486-2497.

[3]   Ford, E. S ; Giles, W. H.; Mokdad, A. H. Increasing prevalence of the metabolic syndrome among U.S. adults. *Diabetes Care,* **2004**, *27*, 2444-2449.

[4]   Cook, S.; Weitzman, M.; Auinger, P.; Nguyen, M.; Dietz, W. H. Prevalence of the metabolic syndrome phenotype in adolescents: findings from the third National Health and Nutrition Examination Survey, 1988-1994. *Arch. Pediatr. Adolesc. Med.*, **2003**, *157*, 821-827.

[5]   Cruz, M. L.; Goran, M. I. The metabolic syndrome in children and adolescents. *Curr. Diab. Rep.*, **2004**, *4*, 53-62.

[6]   de Ferranti, S. D.; Gauvreau, K.; Ludwig, D. S.; Neufeld, E. J.; Newburger, J. W.; Rifai, N. Prevalence of the metabolic syndrome in American adolescents: findings from the Third National Health and Nutrition Examination Survey. *Circulation*, **2004**, *110*, 2494-2497.

[7]   Denys, K.; Cankurtaran, M.; Janssens, W.; Petrovic, M. Metabolic syndrome in the elderly: an overview of the evidence. *Acta Clin. Belg.*, **2009**, *64*, 23-24.

[8]   Eckel, R. H.; Grundy, S. M.; Zimmet, P. Z. The metabolic syndrome. *Lancet,* **2005**, *365*, 1415-1428.

[9]   Prichard, N. B.; Smith, C. C. T.; Ling, K. L.; Betteridge, D. J. Fish oil and cardiovascular disease. *BMJ,* **1995**, *310*, 819-820.

[10]   Weldon, S. M.; Mullen, A. C.; Loscher, C. E.; Hurley, L. A.; Roche, H. M. Docosahexaenoic acid induces an anti-inflammatory profile in lipopolysaccharide-stimulated human THP-1 macrophages more effectively than eicosapentaenoic acid. *J. Nutr. Biochem.*, **2007**, *18*, 250-258.

[11]   Jung, U. J.; Torrejon, C.; Tighe, A. P.; Deckelbaum, R. J. n-3 Fatty acids and cardiovascular disease: mechanisms underlying beneficial effects. *Am. J. Clin. Nutr.,* **2008**, *87*, 2003S-2009S.

[12]   Bloomer, R. J.; Larson, D. E.; Fisher-Wellman, K. H.; Galpin, A. J.; Schilling, B. K. Effect of eicosapentaenoic and docosahexaenoic acid on resting and exercise-induced inflammatory and oxidative stress biomarkers: a randomized, placebo controlled, cross-over study. *Lipids Health Dis.,* **2009**, *8*, 36. doi:10.1186/1476-511X-8-36.

[13]   Blackburn, G. L. From bench to bedside: novel mechanisms and therapeutic advances through the development of selective peroxisome proliferator-activated receptor γ modulators. *Am. J. Clin. Nutr.*, **2010**, *91*, 251S-253S.

[14]   Trayhurn, P.; Wood, I. S. Adipokines: inflammation and the pleiotropic role of white adipose tissue. *Br. J. Nutr.*, **2004**, *92*, 347-355.

[15]   Unger, R. H. Lipotoxic diseases. *Annu. Rev. Med.*, **2002**, *53*, 319-336.

[16]   Kougias, P.; Chai, H.; Lin, P. H.; Yao, Q.; Lumsden, A. B.; Chen, C. Effects of adipocyte-derived cytokines on endothelial functions: implications of vascular disease. *Surg. Res.*, **2005**, *126*, 121-129.

[17]   Kang, J. X.; Weylandt, K. H. Modulation of inflammatory cytokines by omega-3 fatty acids. *Subcell Biochem.*, **2008**, *49*, 133-143.

[18]   Calder, P. C. Immunomodulation by omega-3 fatty acids. *Prostaglandin Leukot. Essent. Fatty Acids*, **2007**, *77*, 327-335.

[19]   Gruppo Italiano per lo Studio della Sopravvivenza nell'Infarto miocardico. Dietary supplementation with n-3 polyunsaturated fatty acids and vitamin E after myocardial infarction: results of the GISSI-Prevenzione trial. *Lancet*, **1999**, *354*, 447-445.

[20]   Fritsche, K. Important differences exist in the dose-response relationship between diet and immune cell fatty acids in humans and rodents. *Lipids*, **2007**, *42*, 961-979.

[21]   Buckley, J. D.; Howe, P. R. Anti-obesity effects of long-chain omega-3 polyunsaturated fatty acids. *Obes. Rev.*, **2009**, *10*, 648-659.

[22]   Li, J. J.; Huang, C. J.; Xie, D. Anti-obesity effects of conjugated linoleic acid, docosahexaenoic acid, and eicosapentaenoic acid. *Mol. Nutr. Food Res.*, **2008**, *52*, 631-645.

[23]   Clarke, S. D. Polyunsaturated fatty acid regulation of gene transcription: a molecular mechanism to improve the metabolic syndrome. *J. Nutr.*, **2001**, *131*, 1129-1132.

[24]   Abeywardena, M. Y.; Head, R. J. Long-chain n-3 polyunsaturated fatty acids and blood vessel function. *Cardiovascular Res.*, **2001**, *52*, 361-371.

[25]   Roos, B.; Mavromatis, Y.; Bronwer, I. A. Long chain n-3 polyunsaturated fatty acids: new insights into mechanisms relating to inflammation and coronary heart disease. *Br. J. Pharmacol.*, **2009**, *158*, 413-428.

[26]   Engler, M. M.; Engler, M. B.; Pierson, D. M.; Molteni, L. B.; Molteni, A. Effects of docosahexaenoic acid on vascular pathology and reactivity in hypertension. *Exp. Biol. Med.*, **2003**, *228*, 299-307.

[27]   Morris, M. C.; Sacks, F.; Rosner, B. Does fish oil, lowers blood pressure? A meta-analysis of controlled trials. *Circulation*, **1993**, *88*, 523-533.

[28]   Geleijnse, J. M.; Giltay, E. J.; Grobbe, D. E.; Donders, A. R.; Kok, F. J. Blood pressure responses to fish oil supplementation: metaregression analysis of randomized trials. *J. Hyperten.*, **2002**, *20*, 1493-1499.

[29]   Engler, M. M.; Engler, M. B.; Goodfriend, T. L.; Ball, D. L.; Yu, Z.; Su, P.; Kroetz, D. L. Docosahexaenoic acid is an antihypertensive nutrient that affects aldosterone production in SHR. *Proc. Soc. Exp. Biol. Med.*, **1999**, *221*, 32-38.

[30]   Mozaffarian, D.; Geelen, A.; Brouwer, I. A.; Geleijnse, J. M.; Zock, P. L.; Katan, M. B. Effect of fish oil on heart rate in humans: a meta-analysis of randomized controlled trials. *Circulation*, **2005**, *112*, 1945-1952.

[31]   Grundy, S. M.; Cleeman, J. I.; Daniels, S. R.; Donato, K. A.; Eckel, R. H.; Franklin, B. A.; Gordon, D. J.; Krauss, R. M.; Savage, P. J.; Smith, S. C.; Spertus, J. A.; Costa, F. American Heart Association; National Heart, Lung, and Blood Institute. Diagnosis and management of the metabolic syndrome: an American Heart Association/National Heart, Lung, and Blood Institute Scientific Statement. *Circulation*, **2005**, *112*, 2735-2752.

[32]   Harris, W. S. n-3 Fatty acids and serum lipoproteins: human studies. *Am. J. Clin. Nutr.,* **1997**, *65*, 1645S-1654S.

[33]   Balk, E. M.; Lichtenstein, A. H.; Chung, M.; Kupelnick, B.; Chew, P.; Lau, J. Effects of omega-3 fatty acids on serum markers of cardiovascular disease risk: a systematic review. *Atherosclerosis*, **2006**, *189*, 19-30.

[34]   Bays, H. Clinical overview of Omacor: a concentrated formulation of omega-3 polyunsaturated fatty acids. *Am. J. Cardiol.*, **2006**, *98*, 71i-76i.

[35]   McKenney, J. M.; Sica, D. Role of prescription omega-3 fatty acids in the treatment of hypertriglyceridemia. *Pharmacotherapy*, **2007**, *27*, 715-728.

[36]   Skulas-Ray, A. C.; West, S. G.; Davidson, M. H.; Kris-Etherton, P. M. Omega-3 fatty acid concentrates in the treatment of moderate hypertriglyceridemia. *Expert. Opin. Pharmacother.*, **2008**, *9*, 1237-1248.

[37]   Huff, M. W.; Telfors, D. E. Dietary fish oil increases the conversion of VLDL apoB to LDL apoB. *Arteriosclerosis*, **1989**, *9*, 58-66.

[38]   Schectman, G.; Kaul, S.; Kissebah, A. H. Heterogeneity of low density lipoprotein responses to fish-oil supplementation in hypertriglyceridemic subjects. *Arteriosclerosis*, **1989**, *9*, 345-354.

[39]   Khan, S.; Minihane, A. M.; Talmud, P. J.; Wright, J. W.; Murphy, M. C.; Williams, C. M.; Griffin, B. A. Dietary long-chain n-3 PUFAs increase LPL gene expression in adipose tissue of subjects with an atherogenic lipoprotein phenotype. *J. Lipid Res.,* **2002**, *43*, 979-985.

[40]   Hilpert, K. F.; West, S. G.; Kris-Etherton, P. M.; Hecker, K. D.; Simpson, N. M.; Alaupovic, P. Postprandial effect of n-3 polyunsaturated fatty acids on apolipoprotein. B-containing lipoproteins and vascular reactivity in type 2 diabetes. *Am. J. Clin. Nutr.*, **2007**, *85*, 369-376.

[41]   Savage, D. B.; Petersen, K. F.; Shulman, G. I. Mechanisms of insulin resistance in humans and possible links with inflammation. *Hypertension*, **2005**, *45*, 828-833.

[42]   Wellen, K. E.; Hotamisligil, G. S. Obesity-induced inflammatory changes in adipose tissue. *J. Clin. Invest.*, **2003**, *112*, 1785-1788.

[43]   Kennedy, A.; Martinez, K.; Chuang, C. C.; LaPoint, K.; McIntosh, M. Saturated fatty acid-mediated inflammation and insulin resistance in adipose tissue: mechanisms of action and implications. *J. Nutr.,* **2009**, *139*, 1-4.

[44]   Reynoso, R.; Salgado, L. M.; Calderon, V. High levels of palmitic acid lead to insulin resistance due to changes in the level of phosphorylation of the insulin receptor and insulin receptor substrate. *Mol. Cell Biochem.*, **2003**, *246*, 155-162.

[45]   Ziemke, F.; Mantzoros, C. S. Adiponectin in insulin resistance: lessons from translational research. *Am. J. Clin. Nutr.*, **2010**, *91*, 258S-261S.

[46]   Gani, O. Are fish oil omega-3 long-chain fatty acids and their derivatives peroxisome proliferator-activated receptor agonists? *Cardiovasc Diabetol.*, **2008**, *7*, 6. Doi:10.1186/1475-2840-7-6.

[47]   Neschen, S.; Morino, K.; Dong, J.; Wang-Fischer, Y.; Cline, G. W.; Romanelli, A. J.; Rossbacher, J. C.; Moore, I. K.; Regittnig, W.; Munoz, D. S.; Kim, J. H.; Shulman, G. I. n-3 Fatty acids preserve insulin sensitivity *in vivo* in a peroxisome proliferator-activated receptor-α-dependent manner. *Diabetes*, **2007**, *56*, 1034-1041.

CHAPTER 5

Effects of n-3 Fatty Acids Supplementation on Insulin Resistance

María Teresa Villarreal-Molina, M.D., Ph.D* and Samuel Canizales-Quinteros, M.Sc., Ph.D*

Unit of Molecular Biology and Genomic Medicine, Instituto Nacional de Ciencias Médicas y Nutrición Salvador Zubirán, Departamento de Biología, Facultad de Química, UNAM.

Abstract: Studies in humans and animal models have demonstrated that fish oil, a natural source of n-3 long chain polyunsaturated fatty acids (LC-PUFA), has clinical significance in the prevention and reversal of insulin resistance. Many studies support that higher proportion of serum and cell-membrane n-3 PUFA can improve insulin sensitivity and prevent type 2 diabetes (T2D). However, epidemiological and dietary intervention studies in healthy, obese and diabetic individuals have produced conflicting results on the metabolic effects of dietary n-3 PUFA. The present chapter discusses updated evidence on this matter, analyzes the possible reasons for these inconsistencies, and the possible mechanisms by which n-3 LC-PUFA may prevent or revert insulin resistance. The use of these fatty acids should be part of integral strategies, considering age, gender, metabolic or health status and other variables. This strategy should include changes in lifestyle, adhering to a healthy diet and regular physical activity. Although this is encouraging in the perspective of insulin resistance prevention, further clinical and basic studies must be designed to confirm and complete our knowledge in this field.

INTRODUCTION

Insulin resistance (IR) is defined as an inadequate response by insulin target tissues, such as skeletal muscle (glucose uptake), liver (inhibition of gluconeogenesis), and adipose tissue (inhibition of lipolysis), to the physiologic effects of circulating insulin [1,2]. Insulin resistance is a major metabolic feature of obesity, and as well as a key factor in the etiology of type 2 diabetes (T2D), generally present for many years before its diagnosis [3].

While approximately 32% of the adult US population is obese, 35% have insulin resistance, and 10% have T2D. In Mexico, the prevalence of IR is unknown but expected to be high, as 30% of the adult population is obese, and 14% have T2D [4]. The prevalence of these diseases has increased sharply over the last two decades, mainly due to sedentary lifestyle and dietary changes. Dietary factors that contribute to the development of IR include high-sucrose and high-fat (saturated and trans-fatty acids) diets [5-7]. In most western countries, 35-40% of total energy ingested is fat, which include various types of fatty acids (FAs) as saturated (SFAs), monounsaturated (MUFA) and polyunsaturated (PUFA) [8]. The latter are classified based on the presence of unsaturated bonds in positions n-6 or n-3. Because observational studies assessing fatty acids composition in serum or tissues suggest that IR is associated with relatively high intake of SFAs (e.g. palmitic acid) and low intake of PUFA (e.g. linoleic acid) [9], nutritionists have focused on dietary characteristics that could contribute to or prevent IR. Various lines of evidence show that specific fatty acids affect cell metabolism, modifying the balance between fatty acids oxidation and lipogenesis. These effects may have important implications in addressing the present epidemic of nutrition-related chronic disease, as increasing n-3 PUFA dietary intake may play a significant role in improving insulin sensitivity [10]. The present chapter reviews updated knowledge on the role of n-3 fatty acids in insulin resistance, and the mechanisms by which n-3 PUFA may prevent or revert insulin resistance.

DIET INTERVENTION STUDIES

Animal Models

The first animal study on the effect of n-6 and n-3 PUFA diet composition was reported by Storlein *et al.* [11], showing that replacement of as little as 6 % n-6 PUFA (safflower oil) with fish oil (rich in n-3 LC-PUFA) was able

---

*Address correspondence to Dr. Samuel Canizales-Quinteros and Dra. M. Teresa Villarreal-Molina:** Unit of Molecular Biology and Genomic Medicine, Instituto Nacional de Ciencias Médicas y Nutrición Salvador Zubirán. Vasco de Quiroga # 15 Colonia Sección 16. Tlalpan 14000. México D.F. Phone and Fax: (00)(52)(55) 56-55-00-11. E-mail: cani@servidor.unam.mx

Maricela Rodríguez-Cruz and Mardia López-Alarcón (Eds)

to prevent the development of IR induced by high-fat feeding in rats. Since then, various studies have established a strong link between dietary lipids, membrane lipids and IR in animal studies. Although the experimental models vary in species (mice or rats), genetic background (wildtype or various knockout models), how insulin resistance is induced (high-fat, high-sucrose, high- fructose, or starch-rich diets), type of intervention (variations in fatty acids composition of experimental diets) and variables measured (fasting plasma glucose (FPG)), homeostasis model of insulin resistance (HOMA-IR), fasting insulin, area under the curve (AUC) glucose, fatty acids composition of the membranes of insulin-sensitive cells, etc.), the overall results from animal studies indicate that dietary n-3 PUFA can both prevent and reverse IR. However, although the n-3 PUFA alpha-linolenic acid (ALA), and LC-PUFA such as eicosapentaenoic acid (EPA) and docosahexaenoic acid (DHA), have all been reported to reduce IR in different animal models, the results have varied in different species and according to the PUFA composition of intervention diets. For instance, while in rats ALA increased FPG as compared to animals supplemented with EPA, DHA, or their mixture; in mice ALA decreased IR. Moreover, in obese rats ALA decreased IR while in non-obese rats it prevented IR and non-alcoholic fatty liver disease (NAFLD) induced by high sucrose diets [12-14]. Likewise, in rats fed with a high fructose diet, ALA alone had no effect on insulin sensitivity and insulin/glucose ratio, while a mix of ALA, EPA and DHA not only prevented decreased insulin sensitivity but greatly improved the strong increase in the n-6:n-3 FA ratio observed in blood serum lipids, cardiac and skeletal muscle membranes induced by the high-fructose diet [15-17].

Human Studies

Multiple epidemiological studies and intervention trials have assessed the role of dietary fat quality on IR. The first report suggesting that the type of fat could influence insulin action in humans was by Kinsell *et al.* [18]. However, although prospective epidemiological studies have reported, a protective effect of fish intake on the development of IR [19, 20], improvement in insulin action as the result of diet n-3 PUFA supplementation in healthy, obese or diabetic individuals has not been consistent.

STUDIES IN HEALTHY INDIVIDUALS

The impact of n-3 PUFA supplementation on metabolic parameters in healthy volunteers has been investigated in recent studies, with inconsistent results. Several studies have found that n-3 PUFA supplementation did not improve insulin sensitivity in healthy men and women: In the OPTILIP study, Griffin *et al.* [21] analyzed healthy men and women aged 40 to 70 years, comparing the effect of diets with different (n-6):(n-3) PUFA ratios where SFA and MUFA concentrations were held constant; (n-6):(n-3) PUFA ratios were adjusted by replacing n-6 PUFA with ALA, EPA, DHA, or a mixture of EPA and DHA. While n-3 PUFA supplementation improved the plasma lipid profile, no differences in fasting plasma glucose (FPG), fasting insulin and HOMA-IR were observed among diets with (n-6):(n-3) PUFA ratios ranging from 2.4 to 11.4. Similarly, Giacco *et al.* [22] studied the effects of moderate fish oil supplementation in healthy men and women comparing two basal diets (one rich in monounsaturated fats and the other rich in saturated fats), where each group was randomized to fish oil supplementation (n-3 PUFA) or placebo. Fish oil did not affect insulin sensitivity, first phase insulin response, beta-cell function or glucose tolerance, suggesting that n-3 PUFA may not improve insulin sensitivity in healthy individuals whose indices are already within the normal ranges.

In contrast, n-3 PUFA have been reported to have an effect in certain populations or age groups. Thorenseng *et al.* [23] measured n-3 fatty acids contents and the n-3/n-6 ratio in erythrocyte membrane phospholipids in the Inuit population of Greenland, finding an inverse association between n-3 (EPA), C22:5 n-3, the n-3/n-6 ratio and HOMA-IR and a positive association between ALA and HOMA-IR. Although this was not a diet intervention study, because the FA composition of erythrocyte membranes are known to mirror the FA pattern of the diet over the past few months, their findings suggest that n-3 PUFA may have a protective effect against IR, although the role of potential confounders such as physical activity, diet, energy intake, socio-economic status and contaminants were not explored. Moreover, in elderly people Tsitouras *et al.* [24] reported that high DHA and EPA consumption increased insulin sensitivity, decreased AUC for glucose and reduced inflammatory markers (serum C reactive protein [CRP] and IL-6). Although this study lacked a concurrent control group and included only 12 individuals, it points out that fish oil supplementation may improve insulin sensitivity in the elderly or other specific populations. In another study, Bloedon *et al.* [25] reported that adding flaxseed to the diets of hypercholesterolemic men and women also improved insulin sensitivity. Participants were randomized to receiving ground flaxseed-containing baked products or matching wheat bran products while following a low fat, low cholesterol diet. The flaxseed diet reduced fasting plasma glucose and the homeostatic model assessment of insulin resistance (HOMA-IR), but did not affect markers of inflammation (IL-6, CRP) or oxidative stress. Although n-3 PUFA may improve insulin sensitivity in certain groups of healthy individuals, n-3 PUFA apparently have no significant effect on healthy individuals with adequate insulin sensitivity.

STUDIES IN OBESE INDIVIDUALS

Various studies have examined the correlation between dietary, plasma or FA composition of different cell types and IR in obese individuals. Micaleff *et al.* [26] reported that obese individuals had significantly lower plasma concentrations of total n-3 PUFA as compared with healthy-weight individuals; and that BMI, waist circumference and hip circumference were inversely correlated with n-3 LC-PUFA, EPA and DHA in obese individuals. In extremely obese individuals, Hernández-Morante *et al.* [27] observed important associations among diet, plasma, and adipose tissue FA composition and various cardiometabolic parameters including plasma insulin levels, While dietary and plasma n-3 PUFA were inversely related to cholesterol, low-density lipoprotein cholesterol (LDL-c), and insulin levels, dietary and adipose tissue trans-fatty-acids and saturated FA (SFA) were associated to increased insulin levels.

On the other hand, several studies have attempted to assess whether PUFA intake improves metabolic parameters including insulin sensitivity in obese or overweight individuals under weight-loss programs, with conflicting results. One of the first studies in obese individuals was reported by Mori *et al.* [28], who analyzed whether dietary fish enhances the effects of weight loss on serum lipids, glucose and insulin in a group of overweight and hypertensive patients. Subjects were randomly assigned to a daily fish meal, an energy-restriction weight-loss diet, or both treatments combined for 16 weeks. The daily fish meal enhanced the decrease in fasting insulin and AUC insulin and glucose obtained by weight loss, although the fish meal alone had no effect on glucose-insulin metabolism. Similarly, Ramel *et al.* [29] conducted a study including 278 overweight to obese men and women (20 to 40 years) who all completed 8 weeks of a energy restriction diet of identical macronutrient composition, but different n-3 LC-PUFA contents. Fish oil supplementation, but not consumption of salmon or lean fish, significantly decreased fasting insulin and HOMA-IR, and this effect was independent of weight loss, triacylglycerol reduction or increased n-3 PUFA in membranes.

In contrast, Ahren *et al.* [30] found no significant difference in fasting levels of glucose, insulin or C-peptide in lean and obese individuals after conjugated linoleic acid (CLA)/n-3 PUFA treatment compared with control oil, and found reduced insulin sensitivity in older obese individuals treated with CLA/n-3 PUFA compared with controls. Similarly, while weight loss improved AUC for glucose and insulin in overweight insulin-resistant women, n-3 PUFA supplementation did not provide added advantage over weight loss alone [31]. The prior inflammatory status of study participants may account for the variance in these results, because another study with obese women showed that fish oil supplementation improved insulin resistance only in those women who were in the top tertile for serum amyloid A [32]. As increased inflammation leads to the development of IR, and n-3 PUFA decreases inflammation, n-3 PUFA may be most effective in subjects with increased inflammation [15].

FISH INTAKE, PUFA AND TYPE 2 DIABETES RISK

Various epidemiological studies have investigated whether there is an association between fish intake and type 2 diabetes risk. Potential benefits of the intake of fish on the development of T2D could be attributed to its high content of dietary n-3 LC-PUFA, specifically EPA and DHA, which could increase insulin sensitivity when present in higher quantities in the cell membranes [33]. Although most epidemiological studies suggest a detrimental role of SFAs and a beneficial role of unsaturated fat on insulin sensitivity [34, 35], overall the results of these studies have also been inconclusive.

Two early cohort studies showed protective effects of fish intake. The study of Feskens *et al.* [36] showed reduced T2D risk for fish eaters compared with non-fish eaters, although the sample was small and had a relatively short follow-up period (4 years). Similarly, in the Dutch and Finnish cohorts of the Seven Countries Study fish consumption was inversely related to 2-h blood glucose levels during a 20-year follow-up of male participants [37]. Studies of the Uppsala Longitudinal Study of Adult Men (ULSAM) cohort consistently suggest that a FA pattern associated with IR is characterized by high proportions of palmitic (16:0), palmitoleic (16:1n-7) and dihomo-γ-linolenic (20:3 n-6) acids and a low proportion of linoleic acid (18 : 2n-6) [38]. A similar FA pattern with high palmitic acid and palmitoleic acid, and low linoleic acid has also independently predicted T2D in the ULSAM and in other cohorts, both prospectively and in cross-sectional studies [38-41]. Moreover, several studies assessing dietary fat intake from food registrations have shown results generally accord with the FA pattern described above, although not consistently.

Epidemiological studies in Iceland are particularly interesting, because despite the high prevalence of obesity, the prevalence of T2D is lower than in other Nordic countries. Thorsdottir *et al.* [42] reported that the prevalence of T2D was inversely associated with both the n-3 PUFA and EPA content of milk, and positively associated with n-6/n-3 FA ratio in milk in Icelandic men. Icelandic milk contains more n-3 PUFA and less n-6 PUFA than milk from other Nordic countries, mainly because animal fodder contains fish meal.

In contrast, in the Nurses Health Study II an association between fish intake and risk of type 2 diabetes was not found [43]. Moreover, prospective studies showed no association. In the Health Professionals Follow-up Study and in the Iowa Women's Health Study, relative T2D risks were 1.01 and 1.1 respectively between the upper and lower quintiles of n-3 LC-PUFA intake [44,45]. Interestingly, a prospective study in older Dutch men and women with a low habitual level of fish intake showed not only that a lack of protection against risk of type 2 diabetes, but a positive association between total fish intake and type 2 diabetes. Because EPA and DHA are present in fatty fish, the fact that lean fish intake accounted for 81% of total fish intake could explain this result. However, in addition to paying attention to the type of fish that is eaten, other components within fish may be related to T2D risk, such as vitamin D which could be negatively associated, and contaminants such as selenium, arsenic and mercury which could be positively associated with T2D [46]. In support of this, serum concentrations of persistent organic pollutants were found to be strongly related with diabetes prevalence in a cross-sectional study. Therefore, particularly at high exposure levels, the potential beneficial effect of EPA and DHA may be counteracted by ingestion of contaminated fish. Finally, differences in range, type, and preparation of fish might explain the observed differences in risk estimates among studies. Deep fat frying can affect the potential benefits of fish by lowering the EPA and DHA content; and the potential beneficial effect of EPA and DHA intake could be counteracted by total cholesterol intake, as elevated cholesterol levels are known to impair pancreatic-cell function and insulin secretion [47].

STUDIES IN T2D PATIENTS

It is generally accepted that supplementation with n-3 PUFA in T2D patients does not have a negative effect on the metabolic parameters, and can improve the serum lipid profile [48].Various studies using low doses of n-3 PUFA ranging from 1 to 2 g/day, have reported no deterioration of glucose control [49-51]. In a recent study conducted in overweight non-insulin treated T2D patients receiving low-fat dietary advice, the group receiving sustained high n-3 PUFA intake through walnut consumption led to a greater reduction in fasting insulin levels than those not receiving walnuts, largely in the first 3 months of dietary treatment [52]. However, several clinical studies have shown adverse effects on blood glucose control and insulin activity in subjects with T2D who consumed large amounts of fish [53-57]. It is now believed that the deleterious effects were largely attributable to the high doses used (e.g. 10 g/day fish oil or more).

The most recent studies reporting the effect of n-3 PUFA diet intervention on IR, insulin sensitivity, and fasting plasma glucose levels in different populations described above are summarized in Table **1**.

Table 1: Effects of n-3 PUFA supplementation on insulin sensitivity in human subjects.

	Subjects	*Design*	*Effect*
Griffin *et al.* [21]	258 healthy men and women aged 45-70 y	Randomized, parallel Comparison of diets with different n-6:n-3 ratios for 6 months.	No difference in FPG insulin or HOMA-IR

Giacco *et al.* [22]	162 healthy men and women	Randomized, parallel Comparison of high SFA diet with or without EPA + DHA for 3 months.	No effect of n-3 PUFA on IVGTT or first phase insulin response
Tsitouras *et al.* [24]	12 healthy men and women aged over 60 years	Longitudinal, diet supplemented with fatty fish + sardine oil or olive + corn oil (placebo), for 8 weeks	Improved insulin sensitivity
Bloedon *et al.* [25]	62 men and postmenopausal women with high cholesterol levels	Randomized, parallel Comparison of diets supplemented with flaxseed or wheat bran for 10 weeks	Improved insulin sensitivity and reduced FPG
Ramel *et al.* [29]	278 obese men and women	Randomized, parallel Energy restriction diet supplemented with lean fish, fatty fish, fish oil or no sea food (control), for 8 weeks	Improved insulin sensitivity and decreased fasting insulin only with fish oil.
Ahren *et al.* [2009]	49 non-diabetic young and old, lean and obese subjects	Randomized, placebo-controlled crossover design Diet supplement with CLA plus n-3 LC-PUFA or control oil for 12 weeks	No effect on FPG or insulin sensitivity in young or in older lean subjects. In older obese subjects, insulin sensitivity was reduced.
Krebs *et al.* [31]	93 overweight insulin-resistant women	Randomized, parallel Weight-loss program with LC n-3 PUFA, weight-loss program with placebo oil, or placebo oil with no weight-loss program (control), for 24 weeks.	No effect of n-3 PUFA on insulin sensitivity. Weight loss improved insulin sensitivity.
Browning *et al.* [32]	30 overweight and obese menopausal non-diabetic women	Randomized, crossover n-3 PUFA supplementation versus placebo compared according to inflammatory status, for 12 weeks.	Improvement in insulin sensitivity only in individuals with raised inflammatory status.
Tapsell *et al.* [52]	50 overweight adults with non-insulin-treated diabetes	Randomized, parallel Low-fat dietary advice, with or without walnut supplement for 1 year.	Greater reductions in fasting insulin levels in the walnut supplementation group.
Mostad *et al.* [56]	26 T2D men and women	Double-blind, controlled Supplement with fish oil or corn oil (placebo) for 9 weeks	Decreased insulin sensitivity and increased FPG with fish oil supplementation.
Barre *et al.* [57]	32 T2D men and women	Randomized, parallel Supplement with flaxseed or safflower (control) oil, for 3 months	No change in fasting glucose or insulin levels

Reasons for Inconsistencies

Both genetic and environmental factors, including lifestyle and diet, contribute to the development of IR. Therefore, variations in any of these factors and the methods used to assess IR can contribute to the inconsistencies between the results of different studies, as previously reviewed by Fedor and Kelley [15].

Health status of the study subjects clearly is a factor that contributes to discrepancies in the results from different studies. The effect of n-3 LC-PUFA will not be evident in healthy individuals who have adequate insulin sensitivity, and n-3 LC-PUFA would not be expected to decrease insulin resistance in diabetic patients who have a combined insulin secretion/insulin resistance defect and may have irreversible damage in insulin-sensitive target tissues. The inconsistencies in results from diet intervention studies in T2D patients have been discussed above. However, it is important to point out that while various studies report that n-3 LC-PUFA can prevent and sometimes reverse the development of insulin, not one study has shown that n-3 LC-PUFA can reverse diabetes.

Both in epidemiological and diet intervention studies, the overall food intake includes a great number of variables that are very difficult to control, including the consumption of antioxidants, micro and macronutrients, the presence of contaminants, the amount and combination of fatty acids consumed, and the duration of dietary intervention, among others. Antioxidants play a very important role because increased oxidative stress increases inflammation, and thus favor IR. Moreover, the amount of dietary fat and the fatty acids/macronutrient composition of the diet also may determine the effectiveness of n-3 LC-PUFA to prevent or reverse insulin resistance, because high-fat or high-sucrose diets rich in saturated, trans, and/or n-6 fatty acids are less likely to respond to supplements with small amounts of n-3 LC- PUFA. In addition, variations in FA dietary components may influence the effect of PUFA, as replacing SFA, TFA, or sucrose with n-6 PUFA may decrease insulin resistance; while replacing n-3 LC-PUFA with n-6 PUFA may worsen insulin resistance because of the anti-inflammatory effects of n-3 PUFA and proinflammatory effects of n-6 PUFA. Variations in EPA and DHA ratios contained in fish oil also contribute to the inconsistencies, because EPA and DHA have different effects on insulin sensitivity. Studies with purified EPA and

DHA and their mixtures in different ratios will help define which n-3 LC-PUFA or which combinations are best in different groups of patients.

MECHANISMS BY WHICH N-3 LC-PUFA MAY PREVENT INSULIN RESISTANCE

Although not fully understood, experimental evidence indicates that various mechanisms explain how n-3 LC-PUFA may reduce or prevent IR.

Fatty Acids Membrane Composition

Dietary fat quality is known to affect the cell membrane fatty acids composition, and consequently cell membrane function [58]. The FA composition of cell membranes may affect several cellular functions, such as the translocation of glucose transporters, altering cell membrane fluidity and ion permeability and by affecting insulin receptor binding/affinity [59]. IR associated with abdominal obesity has been correlated to specific FA patterns in plasma and target tissues of insulin action (liver, fat, and muscle) characterized by increased arachidonic acid (AA) and reduced LA content in rats [60,61] and in humans [62]. In a recent study in a non-obese rat model, insulin resistance was induced with a high fructose diet. This diet induced a strong increase in the (n-6):(n-3) FA ratio of blood serum lipids as well as in cardiac and skeletal muscle membranes, however this increase was of less magnitude in rats fed with ALA, and this improvement was clearly potentiated in rats fed with ALA, EPA and DHA [63], showing that dietary n-3 PUFA can in fact change the cell membrane composition of some insulin-sensitive cell types.

Inflammation

Increased inflammation is one of the major factors leading to the development of insulin resistance. Culp *et al.* [64] were the first to identify that n-3 PUFA reduce lypoxygenase and cyclooxygenase activities and the formation of n-6-PUFA-derived eicosanoids. Both DHA and EPA are known to decrease the release of arachidonic acid by inhibiting phospholipase-2 [65]. In addition, oxygenated metabolites of EPA and DHA formed by COX-2 in the presence of aspirin (called resolvins E and D, respectively) were both found to oppose the effects of inflammatory prostaglandins [66]. Moreover, a lipidomic analysis of insulin-resistant ob/ob mice showed that n-3 LC-PUFA not only inhibited the formation of n-6-PUFA-derived eicosanoids, but more importantly triggered the increased biosynthesis of bioactive intermediaries from n-3 LC-PUFA such as 17-HDHA, resolvin D1 and protectin D1. Resolvins and protectins are a novel series of lipid mediators known to display potent protective actions in experimental colitis, peritonitis, brain ischemia-reperfusion and corneal injury [67-70]. These novel oxygenated products evoke potent protective actions related to the resolution of unremitting inflammation. Thus, the resolution of local inflammation by n-3 LC-PUFA and their derived lipid mediators is likely to underlie the beneficial effects of these compounds on insulin resistance in the mouse model.

In addition, LC-PUFA are known to have an effect on toll like receptors (TLR), which activate the NF-kappaB pathway for the production of inflammatory cytokines when stimulated by bacterial cell wall lipopolysaccharides. Lee *et al.* [71] demonstrated that SFA activate and n-3 LC-PUFA inhibit TLR-2 and TLR-4, and that among the n-3 PUFA, DHA was the most potent inhibitor of this pathway. Moreover, excess glucose and SFAs were recently found to regulate chemotactic factor expression by a mechanism that involves reactive oxygen species (ROS) generation, NFkappaB and PPARgamma, which is repressed by PUFA, where certain SFA triggered chemotactic factor expression via a TLR4-dependent pathway [72].

A very recent study in a transgenic mouse model increased endogenous production of omega-3 LC-PUFA in pancreatic beta-cells through overexpressing a C. elegans fatty acids desaturase gene, *mfat-1*. Interestingly, the cellular increase of n-3 and reduction of n-6 PUFA not only enhanced glucose-, amino acid- and GLP-1-stimulated insulin secretion, the transgenic islets challenged with TNF-alpha, IL-beta and IFN-gamma completely resisted cytokine induced cell death. Cytokine-induced activation of NF-kappaB and extracellular signal-regulated kinase (ERK) was significantly attenuated as a result of n-3 PUFA production. This study is particularly interesting, because n-3 PUFA may be useful to prevent beta-cell death in both type 1 and type 2 diabetes [73].

Adipokines

Another mechanism whereby n-3 LC-PUFA may influence insulin sensitivity is by modifying the profile of adipokines secreted from adipose tissue. Serum levels of several adipokines (adiponectin, leptin, resistin, and visfatin) and are known to contribute to the development of IR, and various studies have examined the effects of n-3 LC-PUFA on the secretion of leptin and adiponectin which exert insulin-sensitizing effects [15].

EPA apparently increases leptin expression. In cultured rat adipocytes, EPA increased both the expression and secretion of leptin [74], and both EPA and DHA partially restored the CLA-induced decrease in plasma leptin in female mice [75]. However, plasma leptin concentrations did not change with fish oil supplementation in overweight male mice fed ad libitum, and leptin concentrations decreased when the mice were fed with a restricted, fish oil-containing diet [76]. In the latter experiments, leptin reduction may have been caused by food restriction and not by fish oil supplementation.

N-3 LC-PUFA also seem to increase adiponectin expression and/or secretion in various models. In the ob/ob mouse model, intraperitoneal injections of DHA induced adiponectin expression in adipocytes both at the mRNA and protein level [77]; dietary DHA, but not EPA supplementation, partially restored the CLA-induced decrease in plasma adiponectin in mice [75], and DHA increased adiponectin expression in porcine myoblasts [78]. Moreover, EPA increased adiponectin secretion, but not adiponectin expression in cultured 3T3-L1 cells, and also increased plasma concentrations in obese human individuals and ob/ob mice [79]. However, Lorente-Cebrain *et al.* [80] reported that EPA decreased adiponectin expression and secretion in cultured rat adipocytes. Altogether, these findings suggest that both EPA and DHA may increase leptin and adiponectin expression and/or secretion in different models.

Gene Expression Regulation

Although there are various mechanisms, n-3 PUFA apparently elicit their effect on IR largely at the genomic level, mainly by regulating gene transcription. Dietary n-3-LC PUFA improve glucose uptake and insulin sensitivity suppressing the expression of lipogenic genes and inducing the expression of genes involved in lipid oxidation, by modulating the activity or expression of various transcription factors where peroxisome proliferator activated receptors (PPAR) play a central role (Fig. **1**), also including sterol regulatory element binding protein-1c (SREBP-1c), hepatic nuclear factors (HNF), retinoid X receptors (RXR), liver X receptor (LXR) [81].

Recent studies have analyzed the effects of fish oils, and EPA, and DHA individually on PPARs and SREBP. PPAR alpha induces fatty acids oxidation and SREBP-1c stimulates fatty acids synthesis. Because fish oil was found to prevent IR induced by high-fat diets in wild type but not in PPAR alpha knockout mice [82]; PPAR alpha is known to mediate the effects of fish oil. Other studies have shown that EPA is a potent activator of PPAR alpha dependent genes, and both EPA and DHA suppress expression of SREBP-1c [83,84]. Moreover, PPAR gamma expression is activated by DHA metabolites accompanied by the increased clearance of glucose in db/db mice and ZDF rats [85]. Thus, n-3 PUFA increase glucose clearance, and fatty acids oxidation, and inhibit fatty acid synthesis through SREBP1c.

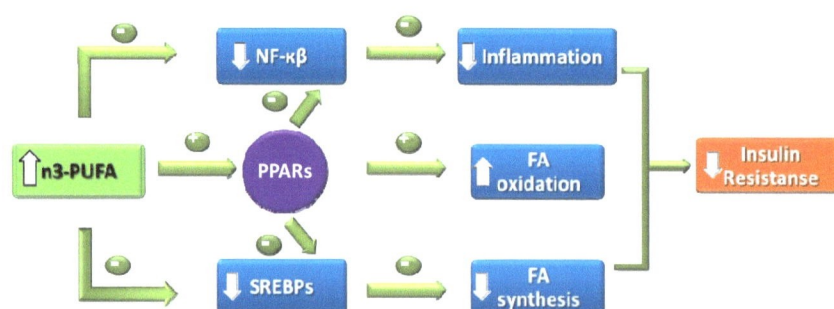

Figure 1: Schematic mechanisms of n-3 PUFA regulation on PPARs and other transcription factors to improve insulin sensitivity.

In conclusion, studies in humans and animal models have demonstrated that fish oil, a natural source of n-3 LC-PUFA such as EPA and AA, has several physiological effects. Many studies support that n-3 LC-PUFA can improve metabolic parameters and prevent T2D. In spite of the concern about a possible deterioration in glucose homeostasis after intake of n-3 fatty acids in patients with T2D, this has generally been observed after using very high doses of n-3 LC-PUFA supplements, and the possible interacting effect of the ingestion of mercury or other contaminants has not been considered. The use of n-3 LC-PUFA should be part of integral strategies, considering age, gender, metabolic or health status and other variables. This strategy should include changes in lifestyle, adhering to a healthy diet and doing regular physical exercise. Although this is encouraging in the perspective of prevention of IR, further clinical and basic studies must be designed to confirm and complete our knowledge in this field. Finally, genetic factors may influence the individual metabolic response to PUFA supplementation, although these genetic factors remain largely unknown [86]. The study of nutrigenomics is just beginning, and will help further define who will benefit more from dietary intervention.

REFERENCES:

[1]   Schenk, S.; Saberi, M.; Olefsky, J.M. Insulin sensitivity: modulation by nutrients and inflammation. *J. Clin. Invest.,* **2008**, *118*, 2992-3002.

[2]   Li, C.; Ford, E.S.; McGuire, L.C.; Mokdad, A.H.; Little, R.R.; Reaven, G.M. Trends in hyperinsulinemia among nondiabetic adults in the U.S. *Diabetes Care,* **2006**, *29*, 2396-2402.

[3]   Zethelius, B.; Berglund, L.; Hänni, A.; Berne, C. The interaction between impaired acute insulin response and insulin resistance predicts type 2 diabetes and impairment of fasting glucose: report from a 20-year follow-up in the Uppsala Longitudinal Study of Adult Men-ULSAM. *J. Med. Sci.,* **2008**, *113*, 117-130.

[4]   Olaiz-Fernández, G.; Rivera-Dommarco, J.; Shamah-Levy, T.; Rojas, R.; Villalpando-Hernández, S.; Hernández-Avila, M. Encuesta Nacional de Salud y Nutrición 2006, Cuernavaca, México. Instituto Nacional de Salud Pública, **2006**.

[5]   Lee, J.S.; Pinnamaneni, S.K.; Eo, S.J.; Cho, I.H.; Pyo, J.H.; Kim, C.K.; Sinclair, A.J.; Febbraio, M.A.; Watt, M.J. Saturated, but not n-6 polyunsaturated, fatty acids induce insulin resistance: role of intramuscular accumulation of lipid metabolites. *J. Appl. Physiol.,* **2006**, *100,* 1467-1474.

[6]   Lovejoy, J.C. The influence of dietary fat on insulin resistance. *Curr. Diab. Rep.,* **2002**, *2*, 435-440.

[7]   Petersen, K.F.; Shulman, G.I. Etiology of insulin resistance. *Am. J. Med.,* **2006**, *119*, S10-16.

[8]   Kinsell, L.W.; Schlierf, G.; Uzawa, H. Dietary considerations with regard to type of fat. *Am. J. Clin. Nutr.,* **1964**, *15*, 198-204.

[9]   Risérus, U. Fatty acids and insulin sensitivity. *Curr. Opin. Clin. Nutr. Metab. Care,* **2008**, *11*, 100-105.

[10] Mann, J.I. Nutrition recommendations for the treatment and prevention of type 2 diabetes and the metabolic syndrome: an evidenced-based review. *Nutr. Rev.,* **2006**, *64*, 422-427.

[11] Storlien, L.H.; Kraegen, E.W.; Chisholm, D.J.; Ford, G.L.; Bruce, D.G.; Pascoe, W.S. Fish oil prevents insulin resistance induced by high-fat feeding in rats. *Science,* **1987**, *237*, 885-888.

[12] Mustad, V.A.; Demichele, S.; Huang, Y.S.; Mika, A.; Lubbers, N.; Berthiaume, N.; Polakowski, J.; Zinker, B. Differential effects of n-3 polyunsaturated fatty acids on metabolic control and vascular reactivity in the type 2 diabetic ob/ob mouse. *Metabolism,* **2006**, *55*, 1365-1374.

[13] Winzell, M.S.; Pacini, G.; Ahrén, B. Insulin secretion after dietary supplementation with conjugated linoleic acids and n-3 polyunsaturated fatty acids in normal and insulin-resistant mice. *Am. J. Physiol. Endocrinol. Metab.,* **2006**, *290*, 347-354.

[14] Murase, T.; Aoki, M.; Tokimitsu, I. Supplementation with alpha-linolenic acid-rich diacylglycerol suppresses fatty liver formation accompanied by an up-regulation of beta-oxidation in Zucker fatty rats. *Biochim. Biophys. Acta,* **2005**, *1733,* 224-231.

[15] Fedor, D.; Kelley, D.S. Prevention of insulin resistance by n-3 polyunsaturated fatty acids. *Curr. Opin. Clin. Nutr. Metab. Care,* **2009**, *12*, 138-146.

[16] Schenk, S.; Saberi, M.; Olefsky, J.M. Insulin sensitivity: modulation by nutrients and inflammation. *J. Clin. Invest.,* **2008**, *118*, 2992-3002.

[17] Robbez-Masson, V.; Lucas, A.; Gueugneau, A.M.; Macaire, J.P.; Paul, J.L.; Grynberg, A.; Rousseau, D. Long-chain (n-3) polyunsaturated fatty acids prevent metabolic and vascular disorders in fructose-fed rats. *J. Nutr.,* **2008**, *138*, 1915-1922.

[18] Kinsell, L.W.; Walker, G.; Michaels, G.D.; Olson, F.E. Dietary fats and the diabetic patient. *N. Engl. J. Med.,* **1959**, *261*:431-434.

[19] Feskens, E.J.; Bowles, C.H.; Kromhout, D. Inverse association between fish intake and risk of glucose intolerance in normoglycemic elderly men and women. *Diabetes Care,* **1991**, *14*, 935-941.

[20] Feskens, E.J.; Virtanen, S.M.; Räsänen, L.; Tuomilehto, J.; Stengård, J.; Pekkanen, J.; Nissinen, A.; Kromhout, D. Dietary factors determining diabetes and impaired glucose tolerance. A 20-year follow-up of the Finnish and Dutch cohorts of the Seven Countries Study. *Diabetes Care,* **1995**, *18*, 1104-1112.

[21]   Griffin, M.D.; Sanders, T.A.; Davies, I.G.; Morgan, L.M.; Millward, D.J.; Lewis, F.; Slaughter, S.; Cooper, J.A.; Miller, G.J.; Griffin, B.A. Effects of altering the ratio of dietary n-6 to n-3 fatty acids on insulin sensitivity, lipoprotein size, and postprandial lipemia in men and postmenopausal women aged 45-70 y: the OPTILIP Study. *Am. J. Clin. Nutr.*, **2006**, *84*, 1290-1298.

[22]   Giacco, R.; Cuomo, V.; Vessby, B.; Uusitupa, M.; Hermansen, K.; Meyer, B.J.; Riccardi, G.; Rivellese, A.A.; KANWU Study Group. Fish oil, insulin sensitivity, insulin secretion and glucose tolerance in healthy people: is there any effect of fish oil supplementation in relation to the type of background diet and habitual dietary intake of n-6 and n-3 fatty acids? *Nutr. Metab. Cardiovasc. Dis.*, **2007**, *17*, 572-580.

[23]   Thorseng, T.; Witte, D.R.; Vistisen, D.; Borch-Johnsen, K.; Bjerregaard, P.; Jorgensen, M.E. The association between n-3 fatty acids in erythrocyte membranes and insulin resistance: The Inuit health in transition study. *Int. J. Circumpolar Health*, **2009**, *68*, 327-336.

[24]   Tsitouras, P.D.; Gucciardo, F.; Salbe, A.D.; Heward, C.; Harman, S.M. High omega-3 fat intake improves insulin sensitivity and reduces CRP and IL6, but does not affect other endocrine axes in healthy older adults. *Horm. Metab. Res.*, **2008**, *40*, 199-205.

[25]   Bloedon, L.T.; Balikai, S.; Chittams, J.; Cunnane, S.C.; Berlin, J.A.; Rader, D.J.; Szapary, P.O. Flaxseed and cardiovascular risk factors: results from a double blind, randomized, controlled clinical trial., *J. Am. Coll. Nutr.*, **2008**, *27*, 65-74.

[26]   Micallef, M.; Munro, I.; Phang, M.; Garg, M. Plasma n-3 Polyunsaturated Fatty Acids are negatively associated with obesity. *Br. J. Nutr.*, **2009**, *102*, 1370-1374.

[27]   Hernández-Morante, J.J.; Larqué, E.; Luján, J.A.; Zamora, S.; Garaulet, M. N-6 from different sources protect from metabolic alterations to obese patients: a factor analysis. *Obesity*, **2009**, *17*, 452-459.

[28]   Mori, T.A.; Bao, D.Q.; Burke, V.; Puddey, I.B.; Watts, G.F.; Beilin, L.J. Dietary fish as a major component of a weight-loss diet: effect on serum lipids, glucose, and insulin metabolism in overweight hypertensive subjects. *Am. J. Clin. Nutr.*, **1999**, *70*, 817-825.

[29]   Ramel. A.; Martinéz, A.; Kiely, M.; Morais, G.; Bandarra, N.M.; Thorsdottir, I. Beneficial effects of long-chain n-3 fatty acids included in an energy-restricted diet on insulin resistance in overweight and obese European young adults. *Diabetologia*, **2008**, *51*, 1261-1268.

[30]   Ahrén, B.; Mari, A.; Fyfe, C.L.; Tsofliou, F.; Sneddon, A.A.; Wahle, K.W.; Winzell, M.S.; Pacini, G.; Williams, L.M. Effects of conjugated linoleic acid plus n-3 polyunsaturated fatty acids on insulin secretion and estimated insulin sensitivity in men. *Eur. J. Clin. Nutr.*, **2009**, *63*, 778-786.

[31]   Krebs, J.D.; Browning, L.M.; McLean, N.K.; Rothwell, J.L.; Mishra, G.D.; Moore, C.S.; Jebb, S.A. Additive benefits of long-chain n-3 polyunsaturated fatty acids and weight-loss in the management of cardiovascular disease risk in overweight hyperinsulinaemic women. *Int. J. Obes.*, **2006**, *30*, 1535-1544.

[32]   Browning, L.M.; Krebs, J.D.; Moore, C.S.; Mishra G.D.; O'Connell, M.A.; Jebb, S.A. The impact of long chain n-3 polyunsaturated fatty acid supplementation on inflammation, insulin sensitivity and CVD risk in a group of overweight women with an inflammatory phenotype. *Diabetes Obes. Metab.*, **2007**, *9*, 70-80.

[33]   Borkman, M.; Storlien, L:H.; Pan, D.A.; Jenkins, A.B.; Chisholm, D.J.; Campbell, L.V. The relation between insulin sensitivity and the fatty-acid composition of skeletal-muscle phospholipids. *N. Engl. J. Med.*, **1993**, *328*, 238-244.

[34]   Lichtenstein, A.H.; Schwab, U.S. Relationship of dietary fat to glucose metabolism. *Atherosclerosis*, **2000**, *150*, 227-243.

[35]   Hu, F.B. The role of n-3 polyunsaturated fatty acids in the prevention and treatment of cardiovascular disease. *Drugs Today*, **2001**, *37*, 49-56.

[36]   Feskens, E.J.; Bowles, C.H.; Kromhout, D. Inverse association between fish intake and risk of glucose intolerance in normoglycemic elderly men and women. *Diabetes Care*, **1991**, *14*, 935-941.

[37]   eskens, E.J.; Virtanen, S.M.; Räsänen, L.; Tuomilehto, J.; Stengård, J.; Pekkanen, J.; Nissinen, A.; Kromhout, D. Dietary factors determining diabetes and impaired glucose tolerance. A 20-year follow-up of the Finnish and Dutch cohorts of the Seven Countries Study. *Diabetes Care*, **1995**, *18*, 1104-1112.

[38]   Vessby, B.; Tengblad, S.; Lithell, H. Insulin sensitivity is related to the fatty acid composition of serum lipids and skeletal muscle phospholipids in 70-year-old men. *Diabetologia*, **1994**, *37*, 1044-1050.

[39]   Laaksonen, D.E.; Lakka, T.A.; Lakka, H.M.; Nyyssönen, K.; Rissanen, T.; Niskanen L.K.; Salonen, J.T. Serum fatty acid composition predicts development of impaired fasting glycaemia and diabetes in middle-aged men. *Diabet. Med.*, **2002**, *19*, 456-464.

[40]   Salomaa, V.; Ahola, I.; Tuomilehto, J.; Aro, A.; Pietinen, P.; Korhonen, H.J.; Penttilä, I. Fatty acid composition of serum cholesterol esters in different degrees of glucose intolerance: a population-based study. *Metabolism*, **1990**, 39, 1285-1291.

[41]   Hodge, A.M.; English, D.R.; O'Dea, K.; Sinclair, A.J.; Makrides, M.; Gibson, R.A.; Giles, G.G. Plasma phospholipid and

dietary fatty acids as predictors of type 2 diabetes: interpreting the role of linoleic acid. *Am. J. Clin. Nutr.*, **2007**, *86*, 189-197.

[42]   Thorsdottir, I.; Hill, J.; Ramel, A. Omega-3 fatty acid supply from milk associates with lower type 2 diabetes in men and coronary heart disease in women. *Prev. Med.*, **2004**, *39*, 630-634.

[43]   Schulze, M.B.; Manson, J.E.; Willett, W.C.; Hu, F.B. Processed meat intake and incidence of type 2 diabetes in younger and middle aged women. *Diabetologia*, **2003**, *46*, 1465-1473.

[44]   van Dam, R.M.; Willett, W.C.; Rimm, E.B.; Stampfer, M.J.; Hu, F.B. Dietary fat and meat intake in relation to risk of type 2 diabetes in men. *Diabetes Care*, **2002**, *25*, 417-424.

[45]   Meyer, K.A.; Kushi, L.H.; Jacobs, D.R.Jr.; Folsom, A.R. Dietary fat and incidence of type 2 diabetes in older Iowa women. *Diabetes Care*, **2001**, *24*, 1528-1535.

[46]   van Woudenbergh, G.J.; van Ballegooijen, A.J.; Kuijsten, A.; Sijbrands, E.J.; van Rooij, F.J.; Geleijnse, J.M.; Hofman, A.; Witteman, J.C.; Feskens E.J. Eating fish and risk of type 2 diabetes: A population-based, prospective follow-up study. *Diabetes Care*, **2009**, *32*, 2021-2026.

[47]   Brunham, L.R.; Kruit, J.K.; Verchere, C.B., Hayden, M.R. Cholesterol in islet dysfunction and type 2 diabetes. *J. Clin. Invest.*, **2008**, *118*, 403-408.

[48]   Dai, J.; Su, Y.X.; Bartell, S.; Le, N.A.; Ling, W.H.; Liang, Y.Q.; Gao, L.; Wu, H.Y.; Veledar, E.; Vaccarino, V. Beneficial effects of designed dietary fatty acid compositions on lipids in triacylglycerol-rich lipoproteins among Chinese patients with type 2 diabetes mellitus. *Metabolism*, **2009**, *58*, 510-518.

[49]   Westerveld, H.T.; de Graaf, J.C.; van Breugel, H.H.; Akkerman, J.W.; Sixma, J.J.; Erkelens, D.W.; Banga, J.D. Effects of low-dose EPA-E on glycemic control, lipid profile, lipoprotein(a), platelet aggregation, viscosity, and platelet and vessel wall interaction in NIDDM. *Diabetes Care*, **1993**, *16*, 683-688.

[50]   Axelrod, L.; Camuso, J., Williams, E.; Kleinman, K.; Briones, E.; Schoenfeld, D. Effects of a small quantity of omega-3 fatty acids on cardiovascular risk factors in NIDDM. A randomized, prospective, double-blind, controlled study. *Diabetes Care*, **1994**, *17*, 37-44.

[51]   Luo, J.; Rizkalla, S.W.; Vidal, H.; Oppert, J.M.; Colas, C.; Boussairi, A.; Guerre-Millo, M.; Chapuis, A.S.; Chevalier, A.; Durand, G.; Slama, G. Moderate intake of n-3 fatty acids for 2 months has no detrimental effect on glucose metabolism and could ameliorate the lipid profile in type 2 diabetic men. Results of a controlled study. *Diabetes Care*, **1998**, *21*, 717-724.

[52]   Tapsell, L.C.; Batterham, M.J.; Teuss, G.; Tan, S.Y.; Dalton, S.; Quick, C.J.; Gillen, L.J.; Charlton, K.E. Long-term effects of increased dietary polyunsaturated fat from walnuts on metabolic parameters in type II diabetes. *Eur. J. Clin. Nutr.,* **2009**, *63*, 1008-1015.

[53]   Sirtori, C.R.; Crepaldi, G.; Manzato, E.; Mancini, M.; Rivellese, A.; Paoletti, R.; Pazzucconi, F.; Pamparana, F.; Stragliotto, E. One-year treatment with ethyl esters of n-3 fatty acids in patients with hypertriglyceridemia and glucose intolerance: reduced triglyceridemia, total cholesterol and increased HDL-C without glycemic alterations. *Atherosclerosis*, **1998**, *137*, 419-427.

[54]   Borkman, M.; Chisholm, D.J.; Furler, S.M.; Storlien, L.H.; Kraegen, E.W.; Simons, L.A.; Chesterman, C.N. Effects of fish oil supplementation on glucose and lipid metabolism in NIDDM. *Diabetes*, **1989**, *38*, 1314-1319.

[55]   Vessby B. n-3 fatty acids and blood glucose control in diabetes mellitus. J. *Intern. Med. Suppl.*, **1989**, *731*, 207-210.

[56]   Mostad, I.L.; Bjerve, K.S.; Bjorgaas, M.R.; Lydersen, S.; Grill, V. Effects of n-3 fatty acids in subjects with type 2 diabetes: reduction of insulin sensitivity and time-dependent alteration from carbohydrate to fat oxidation. *Am. J. Clin. Nutr.*, **2006**, *84*, 540-850.

[57]   Barre, D.E.; Mizier-Barre, K.A.; Griscti, O.; Hafez, K. High dose flaxseed oil supplementation may affect fasting blood serum glucose management in human type 2 diabetics. *J. Oleo. Sci.*, **2008,** *57*, 269-273.

[58] |

[59]   Storlien, L.H.; Pan, D.A.; Kriketos, A.D.; O'Connor, J.; Caterson, I.D.; Cooney, G.J.; Jenkins, A.B.; Baur, L.A. Skeletal muscle membrane lipids and insulin resistance. *Lipids*, **1996**, *31*, S261-265.

[60]   Ginsberg, B.H.; Brown, T.J.; Simon, I.; Spector, A.A. Effect of the membrane lipid environment on the properties of insulin receptors. *Diabetes*, **1981**, *30*, 773-780.

[61]   Fukuchi, S.; Hamaguchi, K.; Seike, M.; Himeno, K.; Sakata, T.; Yoshimatsu, H. Role of fatty acid composition in the development of metabolic disorders in sucrose-induced obese rats. *Exp. Biol. Med.*, **2004**, *229*, 486-493.

[62]   Wilkes, J.J.; Bonen, A.; Bell, R.C. A modified high-fat diet induces insulin resistance in rat skeletal muscle but not adipocytes. *Am. J. Physiol.*, **1998**, *275*, 679-686.

[63]   Warensjo, E.; Riserus, U.; Vessby, B. Fatty acid composition of serum lipids predicts the development of the metabolic syndrome in men. *Diabetologia*, **2005**, *48*, 1999–2005.

[64] Robbez-Masson, V.; Lucas, A.; Gueugneau, A.M.; Macaire, J.P.; Paul, J.L.; Grynberg, A.; Rousseau, D. Long-chain (n-3) polyunsaturated fatty acids prevent metabolic and vascular disorders in fructose-fed rats. *J. Nutr.,* **2008,** *138,* 1915-1922.

[65] Culp, B.R.; Titus, B.G.; Lands, W.E. Inhibition of prostaglandin biosynthesis by eicosapentaenoic acid. *Prostaglandins Med.,* **1979,** *3,* 269-278.

[66] Martin, R.E. Docosahexaenoic acid decreases phospholipase A2 activity in the neurites/nerve growth cones of PC12 cells. *J. Neurosci. Res.,* **1998,** *54,* 805-813.

[67] Serhan, C.N. Systems approach with inflammatory exudates uncovers novel anti-inflammatory and pro-resolving mediators. *Prostaglandins Leukot. Essent. Fatty Acids,* **2008,** *79,* 157-163.

[68] Arita, M.; Yoshida, M.; Hong, S.; Tjonahen, E.; Glickman, J.N.; Petasis, N.A.; Blumberg, R.S.; Serhan, C.N. Resolvin E1, an endogenous lipid mediator derived from omega-3 eicosapentaenoic acid, protects against 2,4,6-trinitrobenzene sulfonic acid-induced colitis, *Proc. Natl. Acad. Sci. U. S. A.,* **2005,** *102,* 7671-7676.

[69] Arita, M.; Bianchini, F.; Aliberti, J.; Sher, A.; Chiang, N.; Hong, S.; Yang, R.; Petasis, N.A.; Serhan, C.N. Stereochemical assignment, antiinflammatory properties, and receptor for the omega-3 lipid mediator resolvin E1. *J. Exp. Med.,* **2005,** *201,* 713-722.

[70] Marcheselli, V.L.; Hong, S.; Lukiw, W.J.; Tian, X.H.; Gronert, K.; Musto, A.; Hardy, M.; Gimenez, J.M.; Chiang, N.; Serhan, C.N.; Bazan, N.G. Novel docosanoids inhibit brain ischemia-reperfusion-mediated leukocyte infiltration and pro-inflammatory gene expression. *J. Biol. Chem.,* **2003,** *278,* 43807-43817.

[71] Gronert, K. Lipoxins in the eye and their role in wound healing. *Prostaglandins Leukot. Essent. Fatty Acids,* **2005,** *73,* 221-229.

[72] Lee, J.Y.; Zhao, L.; Youn, H.S.; Weatherill, A.R.; Tapping, R.; Feng, L.; Lee, W.H.; Fitzgerald, K.A.; Hwang, D.H. Saturated fatty acid activates but polyunsaturated fatty acid inhibits Toll-like receptor 2 dimerized with Toll-like receptor 6 or 1. *J. Biol. Chem.,* **2004,** *279,* 16971-16979.

[73] Han, C.Y.; Kargi, A.Y.; Omer, M.; Chan, C.K.; Wabitsch, M.; O'Brien, K.D.; Wight, T.N.; Chait, A. Differential effect of saturated and unsaturated free fatty acids on the generation of monocyte adhesion and chemotactic factors by adipocytes: dissociation of adipocyte hypertrophy from inflammation. *Diabetes.* **2009,** *59.* 386-396.

[74] Wei, D.; Li, J.; Shen, M.; Jia, W.; Chen, N.; Chen, T.; Su, D.; Tian, H.; Zheng, S.; Dai, Y.; Zhao, A. Cellular production of n-3 PUFAs and reduction of n-6-to-n-3 ratios in the pancreatic beta-cells and islets enhance insulin secretion and confer protection against cytokine-induced cell death. *Diabetes,* **2010,** *59.* 471-478.

[75] Pérez-Matute, P.; Marti, A.; Martínez, J.A.; Fernández-Otero, M.P.; Stanhope, K.L.; Havel, P.J.; Moreno-Aliaga, M.J. Conjugated linoleic acid inhibits glucose metabolism, leptin and adiponectin secretion in primary cultured rat adipocytes. *Mol. Cell. Endocrinol.,* **2007,** *268,* 50-58.

[76] Vemuri, M.; Kelley, D.S.; Mackey, B.E.; Rasooly, R.; Bartolini, G. Docosahexaenoic Acid (DHA) But Not Eicosapentaenoic Acid (EPA) Prevents Trans-10, Cis-12 Conjugated Linoleic Acid (CLA)-Induced Insulin Resistance in Mice. *Metab. Syndr. Relat. Disord.,* **2007,** *5,* 315-322.

[77] Flachs, P.; Mohamed-Ali, V.; Horakova, O.; Rossmeisl, M.; Hosseinzadeh-Attar, M.J.; Hensler, M.; Ruzickova, J.; Kopecky, J. Polyunsaturated fatty acids of marine origin induce adiponectin in mice fed a high-fat diet. *Diabetologia,* **2006,** *49,* 394-397.

[78] González-Périz, A.; Horrillo, R.; Ferré, N.; Gronert, K.; Dong, B.; Morán-Salvador, E.; Titos, E.; Martínez-Clemente, M.; López-Parra, M.; Arroyo, V.; Clària, J. Obesity-induced insulin resistance and hepatic steatosis are alleviated by omega-3 fatty acids: a role for resolvins and protectins. *FASEB J.,* **2009,** *23,* 1946-1957.

[79] Yu, Y.H.; Lin, E.C.; Wu, S.C.; Cheng, W.T.; Mersmann, H.J.; Wang, P.H.; Ding, S.T. Docosahexaenoic acid regulates adipogenic genes in myoblasts via porcine peroxisome proliferator-activated receptor gamma. *J. Anim. Sci.,* **2008,** *86,* 3385-3392.

[80] Itoh, M.; Suganami, T.; Satoh, N.; Tanimoto-Koyama, K.; Yuan, X.; Tanaka, M.; Kawano, H.; Yano, T.; Aoe, S.; Takeya, M.; Shimatsu, A.; Kuzuya, H.; Kamei, Y.; Ogawa, Y. Increased adiponectin secretion by highly purified eicosapentaenoic acid in rodent models of obesity and human obese subjects. *Arterioscler. Thromb. Vasc. Biol.,* **2007,** *27,* 1918-1925.

[81] Lorente-Cebrián, S.; Pérez-Matute, P.; Martínez, J.A.; Marti, A.; Moreno-Aliaga, M.J. Effects of eicosapentaenoic acid (EPA) on adiponectin gene expression and secretion in primary cultured rat adipocytes. *J. Physiol. Biochem.,* **2006,** *62,* 61-69.

[82] Jump, D.B.; Botolin, D.; Wang, Y.; Xu, J.; Christian, B.; Demeure, O. Fatty acid regulation of hepatic gene transcription. *J. Nutr.,* **2005,** *135,* 2503-2506.

[83] Neschen, S.; Morino, K.; Rossbacher, J.C.; Pongratz, R.L.; Cline, G.W.; Sono, S.; Gillum, M.; Shulman, G.I. Fish oil regulates adiponectin secretion by a peroxisome proliferator-activated receptor-gamma-dependent mechanism in mice. *Diabetes,* **2006,** *55,* 924-928.

[84]   Botolin, D.; Wang, Y.; Christian, B.; Jump, D.B. Docosahexaneoic acid (22:6,n-3) regulates rat hepatocyte SREBP-1 nuclear abundance by Erk- and 26S proteasome-dependent pathways. *J. Lipid. Res.,* **2006**, *47,* 181-192

[85]   Shimano, H.; Amemiya-Kudo, M.; Takahashi, A.; Kato, T.; Ishikawa, M.; Yamada, N. Sterol regulatory element-binding protein-1c and pancreatic beta-cell dysfunction. *Diabetes Obes. Metab.,* **2007**, 133-139.

[86]   Yamamoto, K.; Itoh, T.; Abe, D., Shimizu, M.; Kanda, T.; Koyama, T.; Nishikawa, M.; Tamai, T.; Ooizumi, H.; Yamada, S. Identification of putative metabolites of docosahexaenoic acid as potent PPARgamma agonists and antidiabetic agents. *Bioorg. Med. Chem. Lett.,* **2005**, *15,* 517-522.

[87]   Ye, S.Q.; Kwiterovich, P. O. Jr. Influence of genetic polymorphisms on responsiveness to dietary fat and cholesterol. *Am. J. Clin. Nutr.,* **2000**, *72,* 1275S-1284S.

CHAPTER 6

Omega-3 Polyunsaturated Fatty Acids (PUFAs) as the "Mind-Body Interface" in Cardiovascular Diseases and Depression

Jane Pei-Chen Chang, M.D.[a,b] and Kuan-Pin Su, M.D., Ph.D. [a,c]

[a]Department of Psychiatry & Mind-Body Interface Laboratory (MBI-Lab),China Medical University Hospital, Taichung, Taiwan, [b]Graduate Institute of Clinical Medical Science, China Medical University Hospital, Taichung, Taiwan and [c]Graduate Institute of Neural and Cognitive Sciences, China Medical University, Taichung, Taiwan.

Abstract: Depression and cardiovascular diseases (CVDs) are highly comorbid diseases, implying common mechanisms interplay between these two. According to recent clinical and basic studies, omega-3 polyunsaturated fatty acids (n-3 PUFAs) seem to be a possible interface. N-3 PUFAs deficiency is associated with dysfunctions of neuronal membrane stability and transmission of neurotransmitters, which might connect to the etiology of mood and cognitive dysfunction of depression. N-3 PUFAs is essential in balancing the immune function by reducing membrane arachidonic acid and prostaglandin E2 synthesis, which have been linked to the somatic manifestations of CVDs and depression. The role of n-3 PUFAs in immunity and neural function further supports the hypothesis of psychoneuroimmunology of depression and CVDs and provides an excellent interface between 'mind' and 'body.' This review is to provide an overview across evidences of the role of n-3 PUFAs in depression and CVD, and to propose possible mechanisms by which they may act at molecular and cellular levels.

INTRODUCTION

"Sung-shin (傷心)," which literally means "heart-breaking" in Mandarin is similar to the metaphor we use when people are upset or being hurt emotionally, we often describe them to have a "broken heart." Despite biological misleading of the metaphor, accumulating evidence from empirical studies reveal that there seems to be a "mind-body interface" linking cardiovascular diseases (CVDs) with depression [1].

Depression and CVDs are two highly comorbid diseases [2]. Based on the evidence from epidemiological data, case-controlled studies, clinical trials, and meta-analytic reviews, omega-3 polyunsaturated fatty acids (n-3 PUFAs) seem to be a biological link between depression and CVDs [1,3]. Specifically, epidemiological studies have observed societies with high consumption of n-3 PUFAs to have lower prevalence of CVDs, as well as lower prevalence of depression [4]. In addition, case-controlled studies revealed lower levels of n-3 PUFAs in both depression [5] and CVDs patient groups [6]. Study results have further implicated n-3 PUFAs as the link between depression with CVDs via inflammation, hypothalamic-adrenal-axis (HPA) axis hyperactivity and other mechanisms such as neurotransmission.

CVDs AND DEPRESSION

Depression has an estimated 10% life prevalence rate in the general population, and with an estimated 17-27% prevalence rate of depression in patients with CVDs [7]. Meanwhile, 20% of patients with CVDs have major depression [7]. Depression has been associated with mechanisms predisposing to CVDs risks, including lower heart rate variability, autonomous nervous system dysfunctioning, and abnormal lipid metabolisms [7,8]. Moreover, depression without any cardiovascular comorbidity has been found to increase the odds ratio (OR) for future CVDs event (OR=4.5) in the 13-year prospective study from Maryland Epidemiological Catchment Area [7], and patients with a diagnosis of major depressive disorder were 3.9 times more likely to die from cardiac causes compared with those without depression at baseline, even after controlling for disease severity and other risk factors [9]. The exact mechanisms interplaying between depression and CVDs are still under investigation; however, many studies have shown bidirectional relationships between the two to be connected mainly via inflammation and HPA axis hyperactivity. Other mechanisms such as endothelial dysfunction and altered autonomic nervous system activity are also possible connections between CVDs and depression with limited study findings.

Address correspondence to Dr. Kuan-Pin Su: Department of Psychiatry, China Medical University Hospital, No.2 Yuh-Der Road, Taichung 404, Taiwan; Tel: 886-4-22052121-Ext.1073. E-mail: cobolsu@gmail.com

Maricela Rodríguez-Cruz and Mardia López-Alarcón (Eds)

Inflammation

Low grade inflammation has been implied as a common mechanism between CVDs and depression. Inflammation is a complex biological response of vascular tissues to harmful stimuli, such as pathogens, damaged cells, or irritants, and it is a protective attempt by the body to remove the injurious stimuli as well as to initiate the healing process for the tissue. The process of inflammation initiates a cascade of biochemical events involving the vascular system and the immune system. Inflammatory process mediators such as arachidonic acids (AA) and its metabolites, prostaglandins (PGs) and leukotrienes (LTs), contribute to diverse circulatory and homeostasis functions [10]. PGs and LTs are highly biologically active and are known to be involved in various pathological processes, such as atherosclerosis, cardiovascular, cerebrovascular, and pro-inflammatory conditions [11]. Patients with depression have an increase in inflammatory biomarkers including PGE2, interleukin-1 (IL-1), IL-6, and IL-12 when compared with healthy controls [11]. Endotoxin (lipopolysaccharide) or IL-1 induced sickness behavior in animal models further supported the role of inflammation in depression [12]. IL-1 has also been associated with depressive symptoms related to the inhibition of hippocampal neurogenesis in animal studies [13]. Depression is often seen in those patients with chronic inflammation, for example in patients with diabetes mellitus [14] and patients with hepatitis C treated with interferon alpha therapy [15]. On the other hand, patients with low-grade inflammation-related diseases, such as atherosclerosis and diabetes mellitus are also at a higher risk for CVDs [16]. Depressive symptoms may accelerate the progression of atherosclerosis by fostering latent or chronic infections and activating inflammatory processes [17,18]. In other words, depressive symptoms could influence immune regulation, or vice versa. Regardless of the initial causal trigger, an increased vulnerability to adverse cardiac outcomes was observed in depressive patients.

Hypothalamic-Adrenal (HPA) Axis Hyperactivity

HPA axis hyperactivity is another possible common mechanism for depression and CVDs. HPA axis, a major part of the neuroendocrine system, controls reactions to stress and regulates many body processes, including digestion, the immune system, mood and emotions, sexuality, and energy storage and expenditure. HPA axis activity and glucocorticoids actions help to keep the body in balance with the inflammatory process and help down-regulate inflammation when the pathogen was eliminated. Specifically, if white blood cells become overly sensitive to the actions of glucocorticoids, they may down regulate inflammation process prematurely, furthermore, if they become resistant to glucocorticoids, the immune repsonse may continue unchecked and exacerbate inflammatory conditions [19]. The reduced sensitivity to glucocorticoid has been found in lymphocytes of depressed patients [20], while in clinical studies, HPA axis hyperactivity with elevated cerebrospinal fluid (CSF) corticotrophin releasing factor (CRF), blunted adrenocorticotropic hormone response to CRF administration, and nonsuppressed cortisol secretion following dexamethasone administration were found in patients with depression [21, 22]. On the other hand, corticosteroid administrations have been known to increase the risk of CVDs by inducing states of hypercholesterolemia, hypertriglyceridemia, and hypertension [23]. In addition, elevated morning plasma cortisol concentrations have been significantly correlated with moderate to severe coronary atherosclerosis in young and middle aged men [24]. HPA axis hyperactivity was also associated with many cardiovascular disease risk factors, including visceral obesity, increased blood pressure, elevated heart rate, and steroid induced diabetes [25]. Patients with chronic heart failures had higher serum cortisol levels [26], and a recent study reported HPA-axis hyperactivity as a mediating factor between depression and increased risk of CVDs death in patients with depression, especially with a relative risk of 2.3 for men [27].

POLYUNSATURATED FATTY ACIDS (PUFAS) IN DEPRESSION AND CVDS

There are two main types of polyunsaturated fatty acids (PUFAs), the omega-6 (n-6) series (cis-linoleic acid [LA,18:2], γ-linolenic acid [GLA, 18:3, n-6], dihomo-GLA [20:3, n-6], arachidonic acid [AA,20:4, n-6]); and the omega-3 (n-3) series (α-linolenic acid [ALA, 18:3], eicosapentaenoic acid [EPA, 20:5, n-3], docosahexaenoic acid [DHA]). N-3 and n-6 PUFAs are important constituents of all cell membranes; they are essential for survival of humans and other mammals, and cannot be synthesized in the body; hence, they have to be obtained from our diet and are, thus, called essential fatty acids [11].

N-3 PUFAs deficiency has been reported in neurological, cardiovascular, cerebrovascular, autoimmune, metabolic diseases, cancers, and recently in depression [28]. N-3 PUFAs deficiency is associated with dysfunctions of neuronal membrane stability and transmission of neurotransmitters, which might connect to the etiology of mood and cognitive dysfunction of depression. N-3 PUFAs is also important in balancing the immune function and physical health by reducing membrane arachidonic acid (AA) and prostaglandin E2 (PGE2) synthesis, which might be linked to the somatic manifestations physical morbidity, such as CVDs in depression.

N-3 PUFAs in Depression

Epidemiological studies have observed societies with high consumption of n-3 PUFAs appear to have lower prevalence rates of depression [4], and case-controlled studies revealed lower levels of n-3 PUFAs in patients with depression [5]. N-3 PUFAs concentration has been shown to become increasingly suboptimal toward the clinical spectrum of depressive symptoms [29], and its level is significantly negatively correlated with the severity of depressive symptoms [30]. N-3 PUFAs effectively treated depressive disorders in recent studies and some studies have shown EPA alone [31] or a combination of EPA and DHA [32] had positive effects for patients with major depressive disorder. N-3 PUFAs improved the 4-month course of illness in patients with bipolar disorder in a preliminary trial [33], and in the further analysis and of the preliminary trial found that n-3 PUFAs are more beneficial in the depressive phase than in the manic phase in patients with bipolar I disorder [34, 35]. In fact, n-3 PUFAs have been found to have antidepressant effects in several placebo-controlled trials [36].

N-3 PUFAs in CVDs

In the line of evidence showing n-3 PUFAs' relationship with CVDs, societies with high consumption of n-3 PUFAs also appear to have lower prevalence of CVDs [4], and CVDs patient groups also have lower levels of n-3 PUFAs as compared to controls [6]. N-3 PUFAs strongly associated with a reduced risk of sudden death among men without evidence of prior cardiovascular disease [37], and level of n-3 PUFAs may act as a prognostic and a diagnostic utility in CVDs risk assessments [38]. Meta-analysis of n-3 PUFAs on 228,864 individuals with CVDs suggested an increase in fish intake, abundant in n-3 PUFAs, was associated with a 20% reduction in the risk of fatal CVDs, and a 10% reduction in total CVDs [39]. In addition, diets of n-3 PUFAs (especially EPA and/or DHA) were shown to decrease risks for CVDs [37]. The cardiovascular benefit of n-3 PUFAs was shown via its lipid modification effects and its effect in decreasing triglyceride levels by 23% in subjects with mild hyperlipidemia [40], and by 12% in healthy subjects [41] with 4g/day EPA. Supplementation with 1g/day DHA alone or in combination with EPA(1,252mg total) resulted in 21.8% reduction of plasma triglyceride levels in hypertriglyceridemia subjects [42]. In Summary, n-3 PUFAs have shown benefits on CVDs and depression with extensive evidences from epidemiological and clinical studies.

Depression, CVDs and Genes Regulating PUFAs Metabolism

The role of n-3 PUFAs as a linkage between CVDs and depression was further supported by the metabolism pathway of PUFAs and its regulating genes. Phospholipase A2 (PLA2) is the key enzyme of the PUFAs metabolism. PLA2 is a large family of enzymes with Ca2+ independent phospholipase A2 (iPLA2) preferentially on DHA metabolism, while cytosolic PLA2 (cPLA2) is preferentially on AA and EPA metabolism [43,44,45]. On the other hand, cyclo-oxygenase 2 (COX2) is another important enzyme in the metabolism of PUFAs. COX2 converts AA to PGE2, and the later affects immune regulation, neuronal function, and signal transduction, which then in turn are associated with brain dysfunction, presented as depression, sickness behaviors and several physical diseases [28].

It was interesting to find genes of two main enzymes for PUFAs metabolism, the *COX2* gene (or prostaglandin-endoperoxidase synthase 2/*PTGS2)*, and the *cPLA2* gene (*PLA2G4A)*, are located on chromosome 1 and share a common regulatory region [46,47,48]. Furthermore, polymorphisms of *COX2* and *PLA2* genes have been associated with the development of CVDs and depression. For example, a functional $G \rightarrow C$ polymorphism located at 765 base pairs upstream from transcription start site *(-765G $\rightarrow$ C)* has been identified in human *COX2* gene [49], and studies have shown that C allele protects against clinical events including myocardial infarction, stroke [50], cerebrovascular ischemia [51]. C allele at 765 site of *COX2* gene was also associated with lower levels of inflammatory markers such as C-reactive protein and IL-6 in cardiovascular, cerebrovascular, and hypercholesterolemic patients [52]. Although the C allele at 765 site of *COX2* gene has not yet been reported to be associated with depression, another *COX2* polymorphism, *rs468308,* has been found to have significant effects on IFN-induced depression and somatic symptoms in chronic HCV patients receiving INF-α-treatment [15]. Furthermore, the G allele of *Ban I* polymorphism on cPLA2 has been shown to increase the risk of developing depression in a Korean population [48] and the risk for depression and physical symptoms among interferon alpha treated HCV patient groups [15]. The association between genetic polymorphisms of genes that regulate PUFAs metabolism and the comorbidity of CVDs and depression may bring a gleam about the role of n-3 PUFAs in bridging these two diseases.

BIOLOGICAL MECHANISMS LINKING N-3 PUFAS TO DEPRESSION AND CVDS

The biological mechanisms via the regulation of inflammation, HPA axis, neurotransmitters and signal transduction by PUFAs helps to explain the role of n-3 PUFAs in depression and CVDs. PUFAs appear to be directly active in some biological functions; meanwhile some biological functions require conversion of PUFAs into eicosanoids and their products. For example, n-6 AA converts to eicosanoids (PGE2 and LTB4) via COX2 and 5-lipooxygenase (5-LO) in the process of inflammation, which might in turn contribute to the development of somatic symptoms in depression and the physical manifestation of CVDs [3]. In addition, n-3 PUFAs connect CVDs and depression via biological mechanisms such as HPA axis hyperactivity and neuronal membrane stability. The interaction of n-3 PUFAs with the two common biological mechanisms, inflammation and HPA axis hyperactivity, shared by CVDs and depression will be discussed in the following section.

Inflammation

N-3 PUFAs' effects in depression and CVDs have been strongly associated with the 'PUFAs-PGE cascade' hypothesis of inflammation. PUFAs and their metabolites, the eicosanoids (PGs, LTs,TXs), play an important role in modulating biological processes related to brain and physical functions. The 'PUFAs- PGE cascade' hypothesis in depression and CVDs has been supported by a large body of evidence, including higher levels of n-6 AA [35] and hyperactivity of PLA2 in patients with depression [53] and higher levels of n-6 AA in CVDs [54], the inhibitory effect on phospholipase A2 (PLA2) activity of mood stabilizers [55], and the antidepressant effect [31] and anti-arrhythmic effect [56] of n-3 PUFAs.

Pro-inflammatory cytokines, such as IL-1, IL-2, and INF-γ, have been extensively reported in their effects on activities of PLA2 and COX2 and levels of n-6 PUFAs, which further supports the 'PUFAs-PGE cascade' hypothesis in inflammation. For example, treatments with IL-1 can induce activation of cPLA2 in human airway smooth muscle cells [57] and rat dorsal root ganglion cells [58], and induce activation of COX2 in human neuroblastoma cell line [59]. Meanwhile, IFN-γ administration increases cPLA mRNA in the human bronchial epithelial cell line [60]. A systemic release of plasma PLA2 has been found in patients receiving IL-2 therapy during first day of IL-2 infusion [61]. Pro-inflammatory cytokines would then induce oxidant stress, which enhances production of free radicals by monocytes, macrophages, and leukocytes [16]. This increased production of proinflammatory cytokines and free radicals are often seen in conditions associated with a systemic, low-grade, inflammation, including type 2 diabetes mellitus, atherosclerosis, hypertension, hyperlipidemia, and metabolic syndrome X, predisposing to risks of CVDs and depression [18] . N-3 PUFAs on the other hand inhibit free radical generation and prevent oxidant stress [11].

Furthermore, the amount and types of PUFAs released in response to inflammation depends on the phospholipids fatty acids composition on cell membrane, which is determined by the dietary intake and the regulatory enzymes. The beneficial effect of fish consumption with high contents of n-3 PUFAs attributes to the displacement of AA from the cell membrane phospholipids and to preferential formation of less proinflammatory PGs and LTs [11]. In summary, the n-6 AA form eicosanoid series, PGE2 and LTB4, has proinflammatory, proaggregatory, and vasconstrictive effects; while the n-3 PUFAs can antagonize n-6 PUFAs and produce oppositional biological effects.

HPA Axis Hyperactivity

HPA axis hyperactivity is another common biological mechanism supporting n-3 PUFAs' role in depression and CVDs. N-3 PUFAs deficiency has been associated with increased CSF corticotrophin releasing hormone (CRH), which contributes further to the HPA axis hyperactivity [62]. In addition, restoration of dietary DHA normalizes exaggerated distress behavior of n-3 PUFAs deficient rats during administration of CRH [63]. The double-blind placebo-controlled clinical trials have also demonstrated the benefit of 'stress protection' with n-3 PUFAs dietary supplements by showing the attenuation of stress-induced increase in aggression and hostility in Japanese students and significantly reduced perceived stress among stressed university staff [64, 65]. HPA axis hyperactivity has been observed in patients with CVDs and has predisposed CVDs patients to higher risks of sudden cardiac deaths due its affect on the heart and the neuroendocrine system [25]. Meanwhile, HPA axis Hyperactivity is associated with an over-activity of multidrug resistance p-glycoprotein (MDR PGP), a membrane steroid transporter in the brain located on blood-brain-barrier, which results in the reduction of glucocorticoids access to the brain and has been found to relate to the neuronal changes associated with depression [66]. Furthermore, lower plasma long chain omega-3 essential fatty acids status was associated with HPA axis hyperactivity and increased feedback inhibition of the HPA axis, since higher plasma concentrations of neuroactive steroids (3α, 5α-tetrahydrodeoxycorticosterone) specifically for counter-regulating HPA axis hyperactivity was observed in depressive patients [67]. In addition, n-3 PUFAs are able to antagonize the action of pro-inflammatory PGE-effect, in turn normalize MDR PGP over-activity [66].

The role of n-3 PUFAs in CVDs and depression can bee further supported by other mechanisms, including increasing heart rate variability via enhancing vagal nerve functioning [56,68], increasing membrane fluidity [69,70] improving microvascular functioning in cardiovascular and central nerve systems [68,71,72], inhibition of protein kinase C [73], suppression of phophatidylinositol-associated second messenger activity [74], inhibition of angiotensin-converting enzyme, and 3-hydroxy-3-methylglutaryl coenzyme A reductase activities, and their competition with AA for enzymatic action and resultant reduction in inflammatory response [11].

CLINICAL APPLICATIONS

Societies with high consumption of fish, a good source of n-3 PUFAs, appear to have a lower prevalence of depression, coronary heart diseases mortality, CVDs mortality, stroke mortality and all-cause mortality [4,62]. CVDs and depression impact profoundly on human health, as the World Health Organization has pointed out that these 2 conditions will be the 2 top leading causes of disability and premature death in established market economies by 2020 [75]. The current treatment for depression and CVDs are not satisfactory. For example, only less than 50% of depressive patients achieve full remission with optimized medical treatment [76], despite the fact that there are more than 40 antidepressants with mechanisms related to serotonin, norepinephrine and/or dopamine available on the market. Furthermore, depression in patients with CVDs is under-recognized and ineffectively treated due to atypical complaints of somatic discomforts from the patients. The PUFAs hypothesis of depression seems to enlighten a path to discover better treatments for the unresolved depression; meanwhile, n-3 PUFAs' effects on immunomodulation, signal transduction, neurotransmission, neuroprotection, cardioprotection, and anti-arrhythmia provide a cardioprotective effect to help prevent sudden cardiac death and treat arrhythmia in CVDs patients [55]. N-3 PUFAs is essential, safe and beneficial for patients with either depression or CVDs. It also serves as a potential mind-body interface between CVDs and depression, and shows promise of new treatment for patients with both two.

REFERENCES

[1]   Chang, J.; Chen, Y.; Su, K. Omega-3 Polyunsaturated Fatty Acids (n-3 PUFAs) in Cardiovascular Diseases (CVDs) and Depression: TheMissing Link?. *Cardio. Psych. and Neurology*, **2009**, *2009*, 1- 6.

[2]   Glassman, A. H.; Bigger, J. T.; Gaffney, M.; Shapiro, P. A.; Swenson, J. R. Onset of major depression associated with acute coronary syndromes: relationship of onset, major depressive disorder history, and episode severity to sertraline benefit, *Arch. Gen. Psychiatry*, **2006**, *63*, 283- 288 .

[3]   Su, K. P. Biological mechanism of antidepressant effect of omega-3 fatty acids: how does fish oil act as a 'mind-body interface?. *Neurosignals*, **2009**, *17*, 144- 152.

[4]   Hibbeln, J. R.; Nieminen, L. R.; Blasbalg, T. L.; Riggs, J. A.; Lands, W. E. Healthy intakes of n-3 and n-6 fatty acids: estimations considering worldwide diversity. *Am. J. Clin. Nutr.*, **2006**, *83*, 1483S-1493S.

[5]   Peet, M.; Murphy, B.; Shay, J.; Horrobin, D. Depletion of omega-3 fatty acid levels in red blood cell membranes of depressive patients. *Biol. Psychiatry*, **1998**, *43*, 315-319.

[6]   Yokoyama, M.; Origasa, H.; Matsuzaki, M.; Matsuzawa, Y.; Saito, Y.; Ishikawa, Y.; Oikawa, S.; Sasaki, J.; Hishida, H.; Itakura, H.; Kita, T.; Kitabatake, A.; Nakaya, N.; Sakata, T.; Shimada, K.; Shirato, K. Effects of eicosapentaenoic acid on major coronary events in hypercholesterolaemic patients (JELIS): a randomised open-label, blinded endpoint analysis. *Lancet*, **2007**, *369*, 1090-1098.

[7]   Kitzlerova, E.; Anders, M. The role of some new factors in the pathophysiology of depression and cardiovascular disease: overview of recent research. *Neuro. Endocrinol. Lett.*, **2007**, *28*, 832-840.

[8]   Glassman, A. H.; Bigger, J. T.; Gaffney, M.; Van Zyl, L. T. Heart rate variability in acute coronary syndrome patients with major depression: influence of sertraline and mood improvement. *Arch. Gen. Psychiatry*, **2007**, *64*, 1025-1031.

[9]   Penninx, B. W.; Beekman, A. T.; Honig, A.; Deeg, D. J.; Schoevers, R. A.; van Eijk, J. T.; van, T. W. Depression and cardiac mortality: results from a community-based longitudinal study. *Arch. Gen. Psychiatry*, **2001**, *58*, 221-227.

[10]   Gerritsen, M. E. Physiological and pathophysiological roles of eicosanoids in the microcirculation. *Cardiovasc. Res.*, **1996**, *32*, 720-732.

[11]   Das, U. N. Essential fatty acids: biochemistry, physiology and pathology. *Biotechnol. J.*, **2006**, *1*, 420-439.

[12]   Hart, B. L., Biological basis of the behavior of sick animals. *Neurosci. Biobehav. Rev.*, **1988**, *12*, 123-137.

[13]   Goshen, I.; Kreisel, T.; Ben-Menachem-Zidon, O.; Licht, T.; Weidenfeld, J.; Ben-Hur, T.; Yirmiya, R. Brain interleukin-1 mediates chronic stress-induced depression in mice via adrenocortical activation and hippocampal neurogenesis suppression. *Mol. Psychiatry*, **2008**, *13*, 717-728.

[14]   Musselman, D. L. Medical Illness and depression: a delicate interplay between biology and brain, *Programs and abstracts of the American Psychiatric Association 156th Annual Meeting May 17-23, 2003*, American Psychiatric Association, San Francisco California, **2003**; pp. S24B.

[15]   Su, K. P.; Huang, S. Y.; Peng, C. Y.; Lai, C. H.; Huang, C. L.; Chen, Y. C. Phospholipase A2 and cyclo-oxygenase 2 genes influence the risk of interferon-alpha-induced depression by regulating polyunsaturated fatty acids levels. *Biological Psychiatry*, **2009**, [In press].

[16]   Van Guilder, G. P.; Hoetzer, G.L.; Greiner, J. J.; Stauffer, B. L.; Desouza, C. A. Influence of metabolic syndrome on biomarkers of oxidative stress and inflammation in obese adults. *Obesity*, **2006**, *14*, 2127-2131.

[17]   Appels, A.; Bar, F. W.; Bar, J.; Bruggeman, C.; de Baets, M. Inflammation, depressive symptomatology, and coronary artery disease. *Psychosom. Med.*, **2000**, *62*, 601-605.

[18]   Lesperance, F.; Frasure-Smith, N.; Theroux, P.; Irwin, M. The association between major depression and levels of soluble intercellular adhesion molecule 1, interleukin-6, and C-reactive protein in patients with recent acute coronary syndromes. *Am. J. Psychiatry*, **2004**, *161*, 271-277.

[19]   DeRijk, R.; Sternberg E. M., Corticosteroid resistance and disease. *Ann. Med.*, **1997**, *29*, 79-82.

[20]   Pariante, C. M.; Miller, A. H., Glucocorticoid receptors in major depression: relevance to pathophysiology and treatment. *Biol. Psychiatry*, **2001**, *49*, 39-404.

[21]   Raadsheer, F. C.; Hoogendijk, W. J.; Stam, F. C.; Tilders, F. J.; Swaab, D. F., Increased numbers of corticotropin-releasing hormone expressing neurons in the hypothalamic paraventricular nucleus of depressed patients. *Neuroendocrinology*, **1994**, *60*, 436-444.

[22]   Nemeroff, C. B.; Widerlov, E.; Bissette, G.; Walleus, H.; Karlsson, I.; Eklund, K.; Kilts, C. D.; Loosen, P. T.; Vale, W. Elevated concentrations of CSF corticotropin-releasing factor-like immunoreactivity in depressed patients. *Science*, **1984**, *226*, 1342-1344.

[23]   Musselman, D. L.; Evans, D. L.; Nemeroff, C. B. The relationship of depression to cardiovascular disease: epidemiology, biology, and treatment. *Arch. Gen. Psychiatry*, **1998**, *55*, 580-592.

[24]   Troxler, R. G.; Sprague, E. A.; Albanese, R. A.; Fuchs, R.; Thompson, A. J. The association of elevated plasma cortisol and early atherosclerosis as demonstrated by coronary angiography. *Atherosclerosis*, **1977**, *26*, 151-162.

[25]   Rosmond, R.; Bjorntorp, P. The hypothalamic-pituitary-adrenal axis activity as a predictor of cardiovascular disease, type 2 diabetes and stroke. *J. Intern. Med.* **2000**, *247*, 188-197.

[26]   Guder, G.; Bauersachs, J.; Frantz, S.; Weismann, D.; Allolio, B.; Ertl, G.; Angermann, C. E.; Stork, S. Complementary and incremental mortality risk prediction by cortisol and aldosterone in chronic heart failure. *Circulation*, **2007**, *115*, 1754-1761.

[27]   Jokinen, J.; Nordstrom, P. HPA axis hyperactivity and cardiovascular mortality in mood disorder inpatients. *J. Affect. Disord.*, **2009**, *116*, 88-92.

[28]   Su, K. P. Mind-body interface: the role of n-3 fatty acids in psychoneuroimmunology, somatic presentation, and medical illness comorbidity of depression. *Asia Pac. J. Clin. Nutr.*, **2008**, *17* (Suppl 1), 151-157.

[29]   Schiepers, O. J.; de Groot, R. H.; Jolles, J.; van Boxtel, M. P. Plasma phospholipid fatty acid status and depressive symptoms: association only present in the clinical range. *J. Affect. Disord.*, **2009**, *118*, 209-214.

[30]   Edwards, R.; Peet, M.; Shay, J.; Horrobin, D. Omega-3 polyunsaturated fatty acid levels in the diet and in red blood cell membranes of depressed patients. *J. Affect. Disord.*, **1998**, *48*, 149-155 .

[31]   Peet, M.; Horrobin, D. F. A dose-ranging study of the effects of ethyl-eicosapentaenoate in patients with ongoing depression despite apparently adequate treatment with standard drugs. *Arch. Gen. Psychiatry*, **2002**, *59*, 913-919 .

[32]   Su, K. P.; Huang, S. Y.; Chiu, C. C.; Shen, W. W. Omega-3 fatty acids in major depressive disorder. A preliminary double-blind, placebo-controlled trial. *Eur. Neuropsychopharmacol.*, **2003**, *13*, 267-271.

[33]   Stoll, A. L.; Severus, W. E.; Freeman, M. P.; Rueter, S.; Zboyan, H. A.; Diamond, E.; Cress, K. K.; Marangell, L. B. Omega 3 fatty acids in bipolar disorder: a preliminary double-blind, placebo-controlled trial. *Arch. Gen. Psychiatry*, **1999**, *56*, 407-412.

[34]   Su, K. P.; Shen, W. W.; Huang, S. Y. Are omega3 fatty acids beneficial in depression but not mania?. *Arch. Gen. Psychiatry*, **2000**, *57*, 716-717.

[35]   Chiu, C. C.; Huang, S. Y.; Chen, C. C.; Su, K. P. Omega-3 fatty acids are more beneficial in the depressive phase than in the manic phase in patients with bipolar I disorder. *J. Clin. Psychiatry*, **2005**, *66*, 1613-1614.

[36]   Lin, P. Y.; Su, K. P. A meta-analytic review of double-blind, placebo-controlled trials of antidepressant efficacy of omega-3 fatty acids. *J. Clin. Psychiatry*, **2007**, *68*, 1056-1061.

[37]   Albert, C. M.; Campos, H.; Stampfer, M. J.; Ridker, P. M.; Manson, J. E.; Willett, W. C.; Ma, J. Blood levels of long-chain n-3 fatty acids and the risk of sudden death. *N. Engl. J. Med.*, **2002**, *346*, 1113-1118.

[38]   Harris, W. S.; Assaad, B.; Poston, W. C. Tissue omega-6/omega-3 fatty acid ratio and risk for coronary artery disease. *Am. J. Cardiol.*, **2006**, *98*, 19i-26i.

[39]   Whelton, S. P.; He, J.; Whelton, P. K.; Muntner, P. Meta-analysis of observational studies on fish intake and coronary heart disease. *Am. J. Cardiol.*, **2004**, *93*, 1119-1123.

[40]   Mori, T. A.; Burke, V.; Puddey, I. B.; Watts, G. F.; O'Neal, D. N.; Best, J. D.; Beilin, L. J. Purified eicosapentaenoic and docosahexaenoic acids have differential effects on serum lipids and lipoproteins, LDL particle size, glucose, and insulin in mildly hyperlipidemic men. *Am. J. Clin. Nutr.*, **2000**, *71*, 1085-1094.

[41]   Grimsgaard, S.; Bonaa, K. H.; Hansen, J. B.; Nordoy, A. Highly purified eicosapentaenoic acid and docosahexaenoic acid in humans have similar triacylglycerol-lowering effects but divergent effects on serum fatty acids. *Am. J. Clin. Nutr.*, **1997**, *66*, 649- 659.

[42]   Schwellenbach, L. J.; Olson, K. L.; McConnell, K. J.; Stolcpart, R. S.; Nash, J. D.; Merenich, J. A. The triglyceride-lowering effects of a modest dose of docosahexaenoic acid alone versus in combination with low dose eicosapentaenoic acid in patients with coronary artery disease and elevated triglycerides. *J. Am. Coll. Nutr.*, **2006**, *25*, 480-485.

[43]   Strokin, M.; Sergeeva, M.; Reiser, G. Role of Ca2+-independent phospholipase A2 and n-3 polyunsaturated fatty acid docosahexaenoic acid in prostanoid production in brain: perspectives for protection in neuroinflammation. *Int. J. Dev. Neurosci.*, **2004**, *22*, 551-557.

[44]   Strokin, M.; Sergeeva, M.; Reiser, G. Docosahexaenoic acid and arachidonic acid release in rat brain astrocytes is mediated by two separate isoforms of phospholipase A2 and is differently regulated by cyclic AMP and Ca2+. *Br. J. Pharmacol.*, **2003**, *139*, 1014-1022.

[45]   Law, M. H.; Cotton, R. G.; Berger, G. E. The role of phospholipases A2 in schizophrenia. *Mol. Psychiatry*, **2006**, *11*, 547-556.

[46]   Miyashita, A.; Crystal, R. G.; Hay, J. G. Identification of a 27 bp 5'-flanking region element responsible for the low level constitutive expression of the human cytosolic phospholipase A2 gene. *Nucleic Acids Res.*, **1995**, *23*, 293-301.

[47]   Tay, A.; Simon, J. S.; Squire, J.; Hamel, K.; Jacob, H. J.; Skorecki, K. Cytosolic phospholipase A2 gene in human and rat: chromosomal localization and polymorphic markers. *Genomics*, **1995**, *26*, 138-141.

[48]   Pae, C. U.; Yu, H. S.; Kim, J. J.; Lee, C. U.; Lee, S. J.; Lee, K. U.; Jun, T. Y.; Paik, I. H.; Serretti, A.; Lee, C. BanI polymorphism of the cytosolic phospholipase A2 gene and mood disorders in the Korean population. *Neuropsychobiology*, **2004**, *49*, 185-188.

[49]   Papafili, A.; Hill, M. R.; Brull, D. J, McAnulty, R. J.; Marshall, R. P.; Humphries, S. E.; Laurent, G. J. Common promoter variant in cyclooxygenase-2 represses gene expression: evidence of role in acute-phase inflammatory response. *Arterioscler. Thromb. Vasc. Biol.*, **2002**, *22*, 1631-1636.

[50]   Cipollone, F.; Toniato, E.; Martinotti, S.; Fazia, M.; Iezzi, A.; Cuccurullo, C.; Pini, B.; Ursi, S.; Vitullo, G.; Averna, M.; Arca, M.; Montali, A.; Campagna, F.; Ucchino, S.; Spigonardo, F.; Taddei, S.; Virdis, A.; Ciabattoni, G.; Notarbartolo, A.; Cuccurullo, F.; Mezzetti, A. A polymorphism in the cyclooxygenase 2 gene as an inherited protective factor against myocardial infarction and stroke. *JAMA*, **2004**, *291*, 2221-2228.

[51]   Colaizzo, D.; Fofi, L.; Tiscia, G.; Guglielmi, R.; Cocomazzi, N.; Prencipe, M.; Margaglione, M.; Toni, D. The COX-2 G/C -765 polymorphism may modulate the occurrence of cerebrovascular ischemia. *Blood Coagul. Fibrinolysis*, **2006**, *17*, 93-96.

[52]   Orbe, J.; Beloqui, O.; Rodriguez, J. A.; Belzunce, M. S.; Roncal, C.; Paramo, J. A. Protective effect of the G-765C COX-2 polymorphism on subclinical atherosclerosis and inflammatory markers in asymptomatic subjects with cardiovascular risk factors, *Clin. Chim. Acta*, **2006**, *368*, 138-143.

[53]   Noponen, M.; Sanfilipo, M.; Samanich, K.; Ryer, H.; Ko, G.; Angrist, B.; Wolkin, A.; Duncan, E.; Rotrosen, J. Elevated PLA2 activity in schizophrenics and other psychiatric patients. *Biol. Psychiatry*, **1993**, *34*, 641-649.

[54]   Russo, G. L. Dietary n-6 and n-3 polyunsaturated fatty acids: from biochemistry to clinical implications in cardiovascular prevention. *Biochem. Pharmacol.*, **2009**, *77*, 937-946.

[55]   Ghelardoni, S.; Tomita, Y. A.; Bell, J. M.; Rapoport, S. I.; Bosetti, F. Chronic carbamazepine selectively downregulates cytosolic phospholipase A2 expression and cyclooxygenase activity in rat brain. *Biol. Psychiatry*, **2004**, *56*, 248-254.

[56]   Shah, A. P.; Ichiuji, A. M.; Han, J. K.; Traina, M.; El-Bialy, A.; Meymandi, S. K.; Wachsner, R. Y. Cardiovascular and endothelial effects of fish oil supplementation in healthy volunteers. *J. Cardiovasc. Pharmacol. Ther.*, **2007**, *12*, 213-219.

[57]   Pascual, R. M.; Carr, E. M.; Seeds, M. C.; Guo, M.; Panettieri, R. A, Jr., Peters, S. P.; Penn, R. B. Regulatory features of interleukin-1beta-mediated prostaglandin E2 synthesis in airway smooth muscle. *Am. J. Physiol. Lung Cell Mol. Physiol.*, **2006**, *290*, L501-L508.

[58]　Morioka, N.; Takeda, K.; Kumagai, K.; Hanada, T.; Ikoma, K.; Hide, I.; Inoue, A.; Nakata, Y. Interleukin-1beta-induced substance P release from rat cultured primary afferent neurons driven by two phospholipase A2 enzymes: secretory type IIA and cytosolic type IV. *J. Neurochem.*, **2002**, *80*, 989-997.

[59]　Hoozemans, J. J.; Veerhuis, R.; Janssen, I.; Rozemuller, A. J.; Eikelenboom. Interleukin-1beta induced cyclooxygenase 2 expression and prostaglandin E2 secretion by human neuroblastoma cells: implications for Alzheimer's disease. *Exp. Gerontol.*, **2001**, *36*, 559-570.

[60]　Wu, T.; Levine, S. J.; Lawrence, M. G.; Logun, C.; Angus, C. W.; Shelhamer, J. H. Interferon-gamma induces the synthesis and activation of cytosolic phospholipase A2. *J. Clin. Invest.*, **1994**, *93*, 571-577.

[61]　Wolbink, G. J.; Schalkwijk, C.; Baars, J. W.; Wagstaff, J, van den, B. H.; Hack, C. E. Therapy with interleukin-2 induces the systemic release of phospholipase-A2. *Cancer Immunol. Immunother.*, **1995**, *41*, 287-292.

[62]　Hibbeln, J. R.; Bissette, G.; Umhau, J. C.; George, D. T. Omega-3 status and cerebrospinal fluid corticotrophin releasing hormone in perpetrators of domestic violence. *Biol. Psychiatry*, **2004**, *56*, 895-897.

[63]　Takeuchi, T.; Iwanaga, M.; Harada, E. Possible regulatory mechanism of DHA-induced anti-stress reaction in rats. *Brain Res.*, **2003**, *964*, 136-143.

[64]　Bradbury, J.; Myers, S. P.; Oliver, C. An adaptogenic role for omega-3 fatty acids in stress; a randomised placebo controlled double blind intervention study (pilot). *Nutr. J.*, **2004**, *3*, 20.

[65]　Hamazaki, T.; Sawazaki, S.; Itomura, M.; Asaoka, E.; Nagao, Y.; Nishimura, N.; Yazawa, K.; Kuwamori, T.; Kobayashi, M. The effect of docosahexaenoic acid on aggression in young adults. A placebo-controlled double-blind study. *J. Clin. Invest.*, **1996**, *97*, 1129-1133.

[66]　Murck, H.; Song, C.; Horrobin, D. F.; Uhr, M. Ethyl-eicosapentaenoate and dexamethasone resistance in therapy-refractory depression. *Int. J. Neuropsychopharmacol.*, **2004**, *7*, 341-349.

[67]　Nieminen, L. R.; Makino, K. K.; Mehta, N.; Virkkunen, M.; Kim, H. Y.; Hibbeln, J. R. Relationship between omega-3 fatty acids and plasma neuroactive steroids in alcoholism, depression and controls. *Prostaglandins Leukot. Essent. Fatty Acids*, **2006**, *75*, 309-314.

[68]　He, K.; Rimm, E. B.; Merchant, A.; Rosner, B. A.; Stampfer, M. J.; Willett, W. C.; Ascherio, A. Fish consumption and risk of stroke in men. *JAMA*, **2002**, *288*, 3130-3136.

[69]　Rosenberg, I. H. Fish -- food to calm the heart. *N. Engl. J. Med.*, **2002**, *346*, 1102-1103.

[70]　Schwalfenberg, G. Omega-3 fatty acids: their beneficial role in cardiovascular health. *Can. Fam. Physician*, **2006**, *52*, 734-740.

[71]　Marchioli, R.; Barzi, F.; Bomba, E.; Chieffo, C.; Di, G. D.; Di, M. R.; Franzosi, M. G.; Geraci, E.; Levantesi, G.; Maggioni, A. P.; Mantini, L.; Marfisi, R. M.; Mastrogiuseppe, G.; Mininni, N.; Nicolosi, G. L.; Santini, M.; Schweiger, C.; Tavazzi, L.; Tognoni, G.; Tucci, C.; Valagussa, F. Early protection against sudden death by n-3 polyunsaturated fatty acids after myocardial infarction: time-course analysis of the results of the Gruppo Italiano per lo Studio della Sopravvivenza nell'Infarto Miocardico (GISSI)-Prevenzione. *Circulation*, **2002**, *105*, 1897-1903.

[72]　Burr, M. L.; Fehily, A. M.; Gilbert, J. F.; Rogers, S.; Holliday, R. M.; Sweetnam, P. M.; Elwood, P. C.; Deadman, N. M. Effects of changes in fat, fish, and fibre intakes on death and myocardial reinfarction: diet and reinfarction trial (DART). *Lancet*, **1989**, *2*, 757-761.

[73]　Seung Kim, H. F.; Weeber, E. J.; Sweat, J. D.; Stoll, A. L.; Marangell, L. B. Inhibitory effects of omega-3 fatty acids on protein kinase C activity in vitro. *Mol. Psychiatry*, **2001**, *6*, 246-248.

[74]　Bazan, N. G. Cell survival matters: docosahexaenoic acid signaling, neuroprotection and photoreceptors. *Trends Neurosci.*, **2006**, *29*, 263-271.

[75]　Murray, C. J.; Lopez, A. D. *The global burden of disease: a comprehensive assessment of mortality and disability from diseases, injuries, and risk factors in 1990 and projected to 2020*, Harvard University Press, Cambridge: Massachusetts, **1996**.

[76]　Berton, O.; Nestler, E. J. New approaches to antidepressant drug discovery: beyond monoamines. *Nat. Rev. Neurosci.*, **2006**, *7*, 137-151.

CHAPTER 7

Clinical Evidence of Beneficial Effects of Omega-3 Fatty Acids in Allergy and Asthma

Blanca Estela Del Río Navarro, MD, Dino Roberto Pietropaolo Cienfuegos, MD, and Marco Antonio Góngora Meléndez, MD

Department of Allergy and Clinical Immunology, Hospital Infantil de México "Federico Gómez" México D.F.

Abstract: The prevalence of allergic diseases has importantly increased in the last years in some parts of the world. The related causes are poorly known. However, it is thought to be associated to a complex interaction of genetic and environmental factors. There's some epidemiological evidence that the reduction in omega-3 long chain polyunsaturated fatty acids (ω-3 LC-PUFAS) intakes with the increased allergy prevalence. At this time, there's not enough evidence to recommend ω-3 LC-PUFAS supplementation, neither in the pregnant mother, nor in the breastfeeding mother, in order to prevent allergic disease development in high risk population (primary prevention). However, current studies are too heterogeneous and consider only ω-3 LC-PUFAS intake at lower doses. We need to develop more studies with high dose of ω-3 LC-PUFAS to elucidate that issue. Similarly, there are no conclusive data about the prevention of allergy development in high risk infants feeding with ω-3 LC-PUFAS supplementation. ω-3 LC-PUFAS consumption may reduce the use of anti-inflammatory drugs in the treatment of asthma, probably because both of them exert their effects, almost in part, through the same molecular actions. There may be a role of lipid mediators derived from ω-3 LC-PUFAS metabolism (lipoxins and resolvins) in the resolution of allergic asthma inflammation. Hence, synergy between ω-3 LC-PUFAS and drugs may take place and represent an adjunctive therapy in asthma (secondary prevention). Placebo-controlled studies with high-quality methodology design are required so as to draw better conclusions, especially with the employment of ω-3 LC-PUFAS at high doses (2-4 grams per day).

INTRODUCTION

Incidence of allergic diseases has extended worldwide, becoming a healthcare problem. In some countries, up to 50 % of the population has allergic sensitization in some age groups [1]. In this context, research on the leading role of environmental and dietary factors in the development of allergic diseases has been undertaken. A change in people´s diet is one of the factors that need to be considered, since epidemiologic and experimental evidence suggests a connection between a decrease in omega-3 long chain anti-inflammatory polyunsaturated fatty acids (ω-3 LC-PUFAS) intake and the risk of presenting allergic diseases. Attention has been attracted to the potential of these nutrients to prevent allergic diseases. Due to the fact that these diseases usually begin at earlier ages, it is necessary to evaluate the role that preventive measures may play, even in uterus. Adding ω-3 LC-PUFAS to the mother's diet may be a potential intervention in order to prevent the development of allergies in the child (specially, prevention of allergy development in high risk infants with first degree allergic family or primary prevention). In addition, there is a significant number of papers studying the potential role of ω-3 LC-PUFAS supplementation in the treatment of patients with already established allergic diseases (prevention of associated complications or secondary prevention).

ARACHIDONIC ACID AND DOCOSAHEXAENOIC ACID BIOSYNTHESIS

In order to explain the potential role of PUFAS and their metabolic products in allergic disease, we review the biosynthesis pathway briefly. Both linoleic acid (LA) and linolenic acid (LNA) are elongated and desaturated to form long chain polyunsaturated fatty acids (LC-PUFAS) in the endoplasmic reticulum microsomes of the liver. Arachidonic acid [$C_{20}H_{32}O_2$ (AA)], the main product of the ω-6 family, is synthesized from LA through alternating sequences of desaturations and elongations dependent on malonyl coenzyme A (CoA). The same metabolic pathway uses LNA as a substrate to produce eicosapentaenoic acid $C_{20}H_{30}O_2$ (EPA) and docosahexaenoic acid $C_{22}H_{32}O_2$ (DHA) which are the main products of the ω-3 family [2] (see Fig. **1**). Enzymes that make desaturations possible are $\Delta 6$ and $\Delta 5$ desaturases ($\Delta 6D$, $\Delta 5D$). The first desaturation is made by $\Delta 6D$ and it is the limiting step in LC-

---

Address correspondence to Dr. Blanca Estela Del Río Navarro: Department of Allergy and Clinical Immunology, Hospital Infantil de México "Federico Gómez" México D.F; E-mail: blancadelrionavarro@gmail.com

Maricela Rodríguez-Cruz and Mardia López-Alarcón (Eds)

PUFAS synthesis [3]. LA and LNA compete for the same Δ5D and Δ6D [4], the latter is considered to be important for DHA synthesis. This competition explains why linoleic acid intake may reduce DHA and other ω-3 PUFAS metabolism products levels [5]. Enzymatic metabolism of fatty acids DHA, AA, γ-linolenic (GL), dihomo-γ-linolenic (DGL) and EPA produces a wide variety of oxidative products known, as a whole, as eicosanoids. Among the eicosanoids we can find leukotrienes, thromboxanes and prostaglandins. It is worth mentioning that the cell secretes them to the interstitial fluid and their biological actions occur on surrounding cells (paracrine). Prostaglandins stimulate smooth muscle contractions including intestine and uterus. They also mediate pain and inflammation in all tissues. Thromboxanes control platelet function, and consequently coagulation. Due to eicosanoids functions, most signs of unsaturated fatty acids (UFA) deficiency are likely to be caused by alteration in eicosanoids metabolism [6]. AA is the precursor of 2-series prostanoids (prostaglandins and thromboxanes) and 4-series leukotrienes, which exert potent pro-inflammatory activities (bronchoconstriction, vascular permeability and mucus secretion enhancement) in specific organs, while EPA and DHA are the precursors of 3-series prostanoids and 5-series leukotrienes, with predominant anti-inflammatory properties [6].

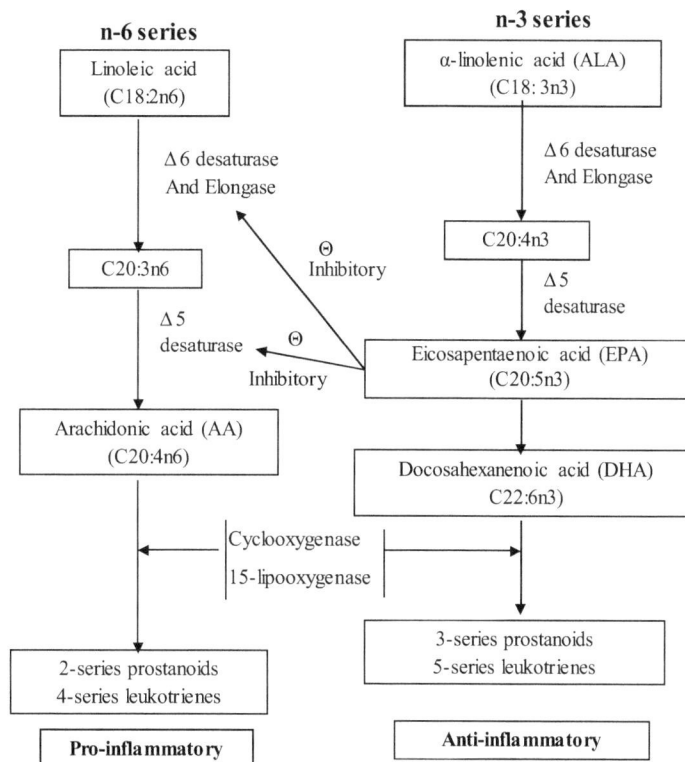

Figure 1: Long chain polyunsaturated fatty acids synthesis from linoleic acid or α-linolenic acid, showing steps catalyzed by Δ5 and Δ6 desaturases. EPA = Eicosapentaenoic acid, DHA = docosahexaenoic acid, AA = arachidonic acid. Adapted from Tawa N. E. and Fisher J. E. In Towsend: Sabiston Textbook of surgery [3].

ALLERGY AND IMMUNE IMBALANCE: CURRENT VIEW

Allergic diseases are generally associated with an inappropriate IgE mediated immune response to ambient allergens that are usually harmless. Conclusive evidence points out that this is due to an underlying T-helper type 2 (Th2) immune cellular response. The process leading to this inappropriate response is not completely known. It was originally suggested that Th2 allergic response occurred as a result of the inhibition of type 1 T-helper (Th1) response in a dual model Th1/Th2 [7]. Subsequently, interest has been directed towards cells that can influence the magnitude and pattern of T-helper cell response. These are called antigen presenting cells (APC), and they provide key signals during the initial encounter with the antigen, and this could potentially have a role on the growing tendency for a Th2 (IL-4 (interleukin-4), IL-5, IL-9, and IL-13) immune response at a population level. Nevertheless, there is no conclusive evidence of a primary defect in APC function. More recent observations have questioned the Th1/Th2 paradigm, which does not explain the simultaneous increase in Th1 (autoimmune diseases) and Th2 (allergic diseases) mediated disorders over the last half of the century [8]. There are also conflicting observations such as a large number of allergic patients having strong Th1 responses to allergens [9,10,11], and Th1 cytokines (IL-2, IFN-γ) playing an important role in the mediation of allergic inflammation both in atopic dermatitis [12] and asthma [13]. A new model that incorporates these observations suggests that changes in modern environment may affect common underlying regulatory pathways, making them less effective in inhibiting inappropriate Th1 and Th2 responses [14]. As a result of this "damage" in immune regulation, the risk of developing both Th2 allergic and Th1 autoimmune diseases increases, depending on individual's genetic predisposition. In consequence, there is an interest in cells with regulatory properties and environmental factors that may have an effect on their function. Dietary ω-3 LC-PUFAS intake appear to be one of the most important changes in environmental factors, and because of its immunomodulatory properties, it represents a subject of interest. Therefore, its potential therapeutic effects are being studied in several inflammatory diseases, such as allergic disease prevention and treatment in early life, when immune response is thought to be in maturing process.

IMPACT IN EARLY LIFE EPISODES AND PREDISPOSITION TO ALLERGIC DISEASES

First events in life seem critical in determining the subsequent kind of immune response and susceptibility to develop allergic diseases. Variations in immune system function in early life have been associated with the increase of susceptibility to inflammatory and allergic diseases [15-19] of the airway [20]. Therefore, environmental factors that potentially influence immune development during this period are important for primary prevention of allergic diseases.

Besides the role of ω-3 LC-PUFAS in fetal neuro-development, these fatty acids may be a key factor for the development of other parts of the body, such as the immune system. During pregnancy, there is immune tolerance between mother and fetus, which is mediated by a complex process at the maternal-fetal interface. Fatty acids, particularly DHA, inhibit cell-mediated immune response because they regulate and decrease major histocompatibility complex (MHC) class II expression [21]. Depletion of these fatty acids, which normally are concentrated in fetal blood (10-15-fold maternal levels), had been associated with pre-term birth [22]. Another significant tolerance mechanism involves the relative predominance of Th2 response that helps in preventing Th1 mediated maternal rejection of fetal graft [23]. This is also associated with changes in fatty acids and prostaglandins metabolism [24]. Nevertheless, although ω-3 LC-PUFAS are known to influence cytokines response, their precise role in early immune programmation still remains under investigation.

This significant modulation mechanism of T cell activity during early development emphasizes the potential effect of maternal ω-3 LC-PUFAS intake during pregnancy for the development of immune response against several proteins, including allergens, at earlier developmental stages. The first allergen-specific response is detected as early as in the 22nd gestation week [25], though this does not seem to be a "typical" memory response [26, 27] and the potential "initiation" mechanism is still controversial [28]. However, allergens and other antigens induce lymphocytic responses in umbilical cord blood. These responses are MHC class II dependent, associated with uncommonly high apoptosis levels [29]. Moreover, surviving cells population seems to possess "suppressing" or "regulating" properties in culture. These cells also express markers including CD4 (cluster of differentiation-4), CD25, and CTLA-4 (cytotoxic T-lymphocyte antigen-4), which are common, though not exclusive, to regulating T cells [29]. The meaning of this remains unclear and more investigation is needed.

An increased level of both ω-6 PUFAS (AA) and ω-3 PUFAS (EPA, DHA) metabolites was found at birth in infants with allergic mothers compared with infants without allergic mothers [30]. In addition, it was showed an altered proportional relationship between the various ω-6 and ω-3 PUFAS metabolites in infants who developed allergic disease during their first 6 years of life. However, other studies found that the relationship between ω-3/ ω-6 LC-PUFAS profile in umbilical cord blood and subsequent allergy is variable and non-predictive [31, 32]. Associations between high ω-3 LC-PUFAS levels in maternal milk and newborn atopy are also inconclusive. Duchen noted that high ω-3 LC-PUFAS levels in maternal milk protect the infant against atopy [33], while Stoney and his collaborators, in a more recent study [34], reached the exact opposite conclusion. Those conflicting results may be related to different study design and at this time we do not know if the discrepancies reflex changes in fatty acids consumption patterns or changes in metabolism of atopic subjects.

ω-3 LC-PUFAS EFFECT ON IMMUNE FUNCTION

ω-3 LC-PUFAS anti-inflammatory properties have been recently revised [35, 36, 37, 38]. As a cell membrane vital structural component, ω-3 LC-PUFAS have the ability to modulate its fluidity and affect cellular signaling. These fatty acids are also precursors for the synthesis of eicosanoids, which have powerful effects on immune response. Eicosanoids derived from arachidonic acid oxidative metabolism are important mediators in the development of asthma in the airway [39]. AA metabolism through cyclooxygenase pathway results in prostaglandins (PG) E_2, D_2, and thromboxanes production, whereas metabolism through 5-lypoxygenase produces leukotrienes (LT) B_4, and cysteinyl leukotrienes (C_4, D_4, E_4). All of these are mediators of bronchoconstriction, vascular permeability, and mucus secretion in specific organs. Supplementation of fish oil (rich in EPA and DHA) reduces the amount of AA available for 4-series LT biosynthesis [40] and other pro-inflammatory mediators since ω-3 and ω-6 LC-PUFAS competes with common desaturation enzymes. It has been likewise showed that DHA and EPA produce lipid anti-inflammatory mediators: resolvins and docosatrienes, which are liberated by neutrophils [41].

Recently, ω-3 LC-PUFAS effect has been reported within the context of inflammatory and autoimmune chronic diseases mediated by Th1 cells, such as rheumatoid arthritis, type I diabetes and inflammatory bowel disease, which has stimulated interest in cell signaling as a common mechanism in such diseases. New data suggest that ω-3 LC-PUFAS membrane composition modifies lipid microdomains (lipid rafts) functionality. These microdomains mediate cellular responses in a variety of tissues [42]. DHA intake from the diet has been associated with kinase C and IL-2 production suppression in murine models [43], which results in increased apoptosis of T CD4+ cells polarized with Th1 profile, these effects could be mediated and modulated by the composition of lipid rafts, which is influenced by ω-3 LC-PUFAS intake [44]. The latter hypothesis appears to explain the immunomodulatory properties showed in murine model of Th1 mediated diseases. More studies are needed to confirm such findings in human diseases.

In a study with human participants ingesting fish oil, a significant suppression in 2-cyclooxygenase expression was noted in peripheral blood monocytes when stimulated *in vitro* with lipopolysaccharides working as agonists of Toll-like receptor 4 and 2 (TLR 4 and 2) [45], suggesting that immunomodulatory effects of ω-3 LC-PUFAS on chronic inflammatory diseases is in part, mediated by TLR pathway. In addition, LC-PUFAS ω-3 role as mediators for these effects has been recently focused on genes expression by micro-assay analyses [46], particularly, on the effect of fatty acids and their binding with at least three nuclear receptors capable to affecting gene transcription: peroxisome proliferator-activated receptor (PPAR-alfa), hepatocyte nuclear factor 4A (HNF4-alpha), liver X receptor (LXR-alpha) and on the regulation of the transcription factor sterol regulatory element binding protein (SREBP1-c) (see Fig. 2) [47]. However, afore mentioned actions had been related clearly with lipogenic gene expression and at this moment we find only indirect evidence implicating anti-inflammatory effect of LC-PUFAS with genetic mechanism, such as suppression of the production of IL-1 messenger ribonucleic acid (mRNA), IL-1 and tumor necrosis factor (TNF) by monocytes, both in healthy volunteers as in humans with arthritis, psoriasis and ulcerative colitis [48]. These cytokines are very important in innate immunity and in the early steps of inflammation. However, Kremer *et al.* only found, in patients with rheumatoid arthritis, significative reduction, both in the number of tender joints, as in the necessity to use nonsteroidal anti-inflammatory drugs, with the supplementation of at least 3 daily grams of DHA and EPA [49].

LINKS BETWEEN ω-3 PUFAS AND ALLERGIC DISEASE

In the last decades, the prevalence of atopy and allergic diseases in different developed countries has increased. The aetiology of such diseases has been related to multifactorial interaction of genetic determinants and environmental factors.

Attention has been particularly focused on nutritional factors as part of the population lifestyle change with special emphasis on variations in the composition of fatty acids intake in general, and in ω-6 and ω-3 LC-PUFAS [50], in particular.

There are controversial results, both in intervention trials, as in studies with cross-sectional design that have examined the links between ω-3 LC-PUFAS and the development of allergic diseases [51]. Systematic reviews have shown beneficial effects of fish oil supplementation in adults with asthma [52] or atopic dermatitis [53], with a lack of data to conclude the effectiveness in atopic children. Other cross-sectional studies in children have shown a strong link between ω-3 LC-PUFAS intake and the risk of an allergic disease. There are data providing evidence, at least in children, that diet rich in ω-3 LC-PUFAS can protect them against symptoms of asthma, permitting an evaluation of the role of early supplementation to prevent the development of allergies [32, 54]. This is the potential role of LC-PUFAS intake in the pregnant mother, in the breastfeeding mother or in the infant during the first months of life, in order to prevent the allergy development, especially in those subjects with first degree siblings or parents with an already established allergic disease (primary prevention of allergic disease).

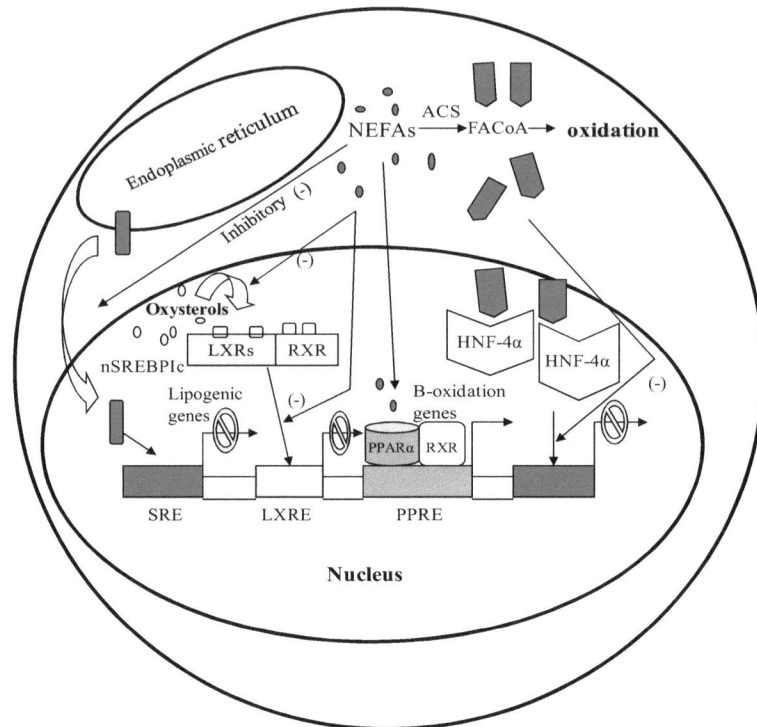

Figure 2: Different mechanisms of gene regulation of lipid metabolism by PUFAS. Direct binding of NEFAs to PPAR results in activation of target genes while binding of FACoA to HNF-4α results in gene repression. NEFAs also inhibit activation of LXRs by oxysterols, and inhibit binding of LXR to LXREs; these actions result in repression of target genes. Fatty acids have also been reported to inhibit nSREBP1c maturation via proteolytic cleavage from the Golgi apparatus, but it is unclear which forms of the fatty acids accomplish this. FACoA = fatty acyl CoA, HNF = hepatocyte nuclear factor, LXR = liver X receptor, LXRE = LXR response element, NEFAs = non-esterified fatty acids, nSREBP1c = nuclear (mature) sterol regulatory element-binding protein. Adapted from Sampat H., *et al.* [47].

ω-3 LC-PUFAS administration and the decrease in environmental allergens (house dust mites) at home can help to prevent the development of allergic sensitization and airway diseases in infants, thus it is suggested a possible protective effect of ω-3 LC-PUFAS administration on the decrease in inflammatory mediators derived from arachidonic acid, such as prostaglandins and thromboxanes. Peat and coworkers conclude that these interventions may play a role in the prevention of allergies development and airway diseases in the infant [55].

Recently, Anandan *et al.* performed a systematic review in literature about the use of ω-3 and ω-6 PUFAS for primary prevention in allergic disease [56] and, contrary to other evidences, they did not find an important role for this supplementation as a strategy of prevention in atopic eczema, asthma, allergic rhinitis or food allergy. However, the afore mentioned study was heterogeneous in terms of differences in time of supplementation, type of intervention (fish oil vs mixture of canola and tuna oils), subjects included (pregnant or breastfeeding mothers vs newborns) and doses of supplementation (high vs low). Therefore, at this time it's not possible to do powerful evidence based recommendations related to the role of primary prevention of ω-3 and ω-6 LC-PUFAS in allergic disease.

ω-3 LC-PUFAS AND ASTHMA

Asthma is one of the most common chronic respiratory diseases worldwide and is estimated to affect approximately 300 million people of all ages and ethnic groups [57]. Asthma prevalence varies widely when compared among countries, even in teenagers and adults whose diagnosis can be quite precise. The International Study of Asthma and Allergies in Childhood phase III (ISAAC) has found a prevalence of asthma symptoms during the last 12 months, of 3-5% in countries such as Indonesia, China, and Greece, and more than 20% in Canada, Australia, New Zealand, and the United Kingdom. In Mexico, according to the same study, the prevalence of wheezing in the last 12 months amounted to 8.7% in 13-14 years old teenagers and to 8.0% in 6-7 years old preschoolers. The prevalence of serious asthmatic symptoms was 4.3% in 13-14 years old teenagers, and 3.3% in 6-7 years old children [58].

Asthma is a chronic inflammatory disease of the airway characterized by wheezing, difficulty to breathe, excessive production of mucus, and cough. Airway remodeling is a characteristic of asthma evolution and is related with airway hyperresponsiveness as a consequence of very different stimuli. One of the most important parts of asthma treatment is pharmacological therapy. However, other non pharmacological interventions, such as dietary manipulation, could reduce the dose of the drug [57].

Changes in dietary intake patterns have been associated with the development of a more sedentary lifestyle. This could have contributed to the increase in asthma over the past few decades. Diet may have an influence in asthma through the intake of allergens, inflammatory mediators, inflammatory mediator precursors (mainly some fatty acids), electrolytes and antioxidants. Anti-inflammatory properties of ω-3 LC-PUFAS are based on their ability to compete as substrates with arachidonic acid (AA) in the formation of pro-inflammatory mediators such as leukotrienes and prostaglandins, and can directly modulate mediator's production by neutrophil and monocytes, chemotactic response and cytokines production (see Fig. **3**) [57].

There is a body of evidence reviewing the potential role of ω-3 LC-PUFAS as an adjunctive therapy in asthma treatment; that is, in secondary prevention of asthma (prevention of related complications when a disease is already established or decrease in the chronic use of drugs with important side effects, such as corticosteroids). A search of the Cochrane Airways Group Specialised Register found 9 randomized controlled trials between 1986 and 2001 showing absence of any effect on Forced Expiratory Volume within the first second (FEV1), maximum expiratory flow (MEF), asthma symptoms, and in drug use for asthma or bronchial hyperresponsiveness. One of the included studies, carried out in children, which combined dietary manipulation with fish oil supplementation, showed an improvement in the MEF and a reduced use of asthma drugs, without adverse events associated with fish oil supplements. Implications for clinical practice of this review, however, indicate that there is not enough evidence to recommend supplementation or modification of asthma patients' diet with ω-3 LC-PUFAS (fish oil) to improve asthma control. Likewise, there is no evidence of associated important adverse effects related to ω-3 LC-PUFAS intake [59], but the lack of favorable results may be due to trials heterogeneity.

Because of the great inconsistency of respiratory tests results, it is impossible to determine whether ω-3 LC-PUFAS have a role as adjuvants in the therapy of children and adults. Therefore, more extensive randomized controlled trials on effects of ω-3 LC-PUFAS at high doses (at least 2-4 grams per day of a combination with EPA and DHA) on pulmonary function parameters or asthma inflammation markers with controlled diet and physical activity are required [60], especially when the clinical efficacy has been related to dose, as in some rheumatoid diseases [49] and metabolic diseases (hypotrigliceridemic effect) [61, 62].

Recently, it has been evaluated the role of lipid mediators in the resolution of airway inflammation in asthmatic patients, such as lipoxins (LXs) derived from AA metabolism and resolvins derived from ω-3 LC-PUFAS mediators [63]. In murine experimental models, it has been showed that resolvin E1 was related to acute airway inflammation resolution by directly suppressing IL-23 and IL-6, which are integral interleukins in the pathogenesis of chronic inflammation, also it has been showed a concomitant reduction in bronchoalveolar lavage fluids population of leukocytes, eosinophils and macrophages [64]. Other researchers thought that this body of evidence is so important to consider resolvin E1 as a potential novel therapeutic approach for human asthma [65].

There is evidence suggesting that a body mass index increase and dietary patterns, especially those related to fat intake, contribute to asthma symptoms, hence dietary changes could help patients gain asthma control and contribute to health general improvement [66].

PROTECTIVE EFFECT OF ω-3 PUFAS ON EXERCISE-INDUCED BRONCHOCONSTRICTION IN ASTHMA

Exercise-induced bronchoconstriction (EIB) refers to the transitory narrowing of the airways which can occur during and following vigorous exercise, resulting in a post-exercise decrement in pulmonary function. Exercise is a powerful trigger for asthma symptoms and may result in asthmatic patients avoiding physical activity, leading to detrimental consequences to their health. Approximately 80% of individuals with asthma and a high prevalence of non-atopic elite athletes have hyperresponsiveness to exercise and experience EIB [67]. There is accumulating evidence that dietary modification potentially reduces the prevalence and incidence of asthma and EIB. In the typical Western diet, ω-6 LC-PUFAS are 20 to 25 times more consumed than ω-3 LC-PUFAS, which results in the release of pro-inflammatory metabolites. Thus, fish oil supplementation is suggested to be a beneficial nonpharmacological intervention for asthmatic subjects with EIB [67] (See Fig. **3**).

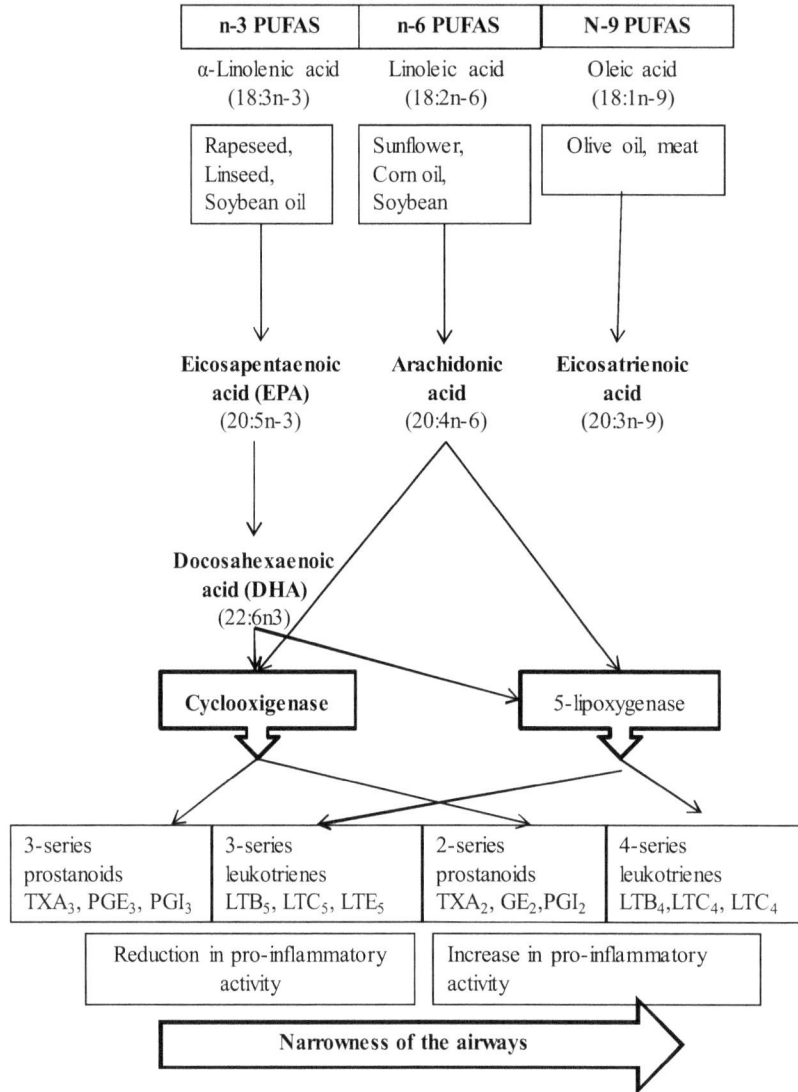

Figure 3: Metabolism of dietary fatty acids after intake through COX and 5-lypooxigenase enzymatic pathways and its effects on inflammatory activity and bronchoconstriction, leukotrienes (LT), prostaglandins (PG) and thromboxanes (TX), Adapted from Mickleborough TD et al [68].

ω-3 PUFAS AND ALLERGIC RHINITIS

Rhinitis refers to the inflammation of the nasal mucosa and is clinically identified by the occurrence of rhinorrhea, nasal itching, sneezing, and nasal congestion. When the etiology is allergic, rhinitis usually occurs with ocular symptoms, such as conjunctival hyperemia, pruritus, edema, and epiphora. This symptomatic response is triggered by environmental aeroallergens [69]. Rhinitis is a global health problem which affects more than 20% of the population in industrialized countries. Its prevalence ranges worldwide from 0.5% to 28% in children, and from 0.5% to 15% in adults [70].

With respect to the association of ω-3 LC-PUFAS intake with allergic diseases, epidemiology is unclear. A birth cohort study realized in Oslo found that fish intake during the first year of life is related to a reduction in risk of allergic rhinitis, yet not of asthma, at the age of 4 years [71].

The relation between allergic rhinitis and ω-3 LC-PUFAS intake has been studied by determining the presence of fatty acids on erythrocytes membrane in order to test the biological availability of fatty acids in cells, as well as its relationship with allergic sensitization and rhinitis. An EPA increase in erythrocytes membrane has been associated with a lower risk of allergic sensitization and clinical symptoms of allergic rhinitis. The consistency of those results reflects a valid pattern of association. Hence, this appears to confirm the hypothesis that consuming ω-3 fatty acids produces favorable results even in adults; although a causal relation has not been established [72] and more studies are required.

One study showed that EPA and DHA intake is associated with a reduction in allergic rhinitis prevalence and that there's no evidence demonstrating that ω-6 LC-PUFAS intake is related to an increase in allergic rhinitis prevalence. There is a trend towards an inverse relation between fish intake and allergic rhinitis, but also without establishing a causal effect [73].

Previous systematic reviews have been considered limited because the available meta-analyses were inappropriate: excessive data loss, poorly defined or heterogeneous study groups, interventions, combined interventions and outcomes. Moreover, these investigations carried out on small samples may make it impossible at this moment to evaluate the impact of clinical outcomes of these co-variables. Finally, because of the lack of significant evidence, it is not possible to evaluate their effectiveness [60].

ω-3 LC-PUFAS AND ATOPIC DERMATITIS

Atopic dermatitis is the most common chronic skin disease in children. In scholars, its prevalence exceeds the 17%. In Mexico, atopic dermatitis symptoms are reported in 6-7 years old children (4.0%) and in 13-14 years old teenagers (28%) [74]. Skin lesions tend to be erythematous and vesicular [75]. In 1933, Hansen found that patients suffering from atopic dermatitis had low concentrations of essential fatty acids (EFAS) in blood. When those patients were fed with corn oil (rich in linoleic acid), skin alterations disappeared [74].

These studies showed that ω-6 EFAS deficiency leads to inflammatory processes in animals' and human beings' skin. Nevertheless, it has been recently proposed that there is not LA deficiency in atopic eczema since there is an increase in LA concentrations in blood, milk, and fatty tissue belonging to patients suffering from atopic eczema, whereas LA metabolites concentrations are abnormally reduced. This demonstrates that there is a reduced transformation from linoleic acid into γ-linolenic acid. Administration of γ-linolenic acid improves skin condition in most of these patients [76]. The majority of investigators who have studied atopic dermatitis biochemistry have concluded that EFAs metabolism in these patients is altered [77].

ω-3 LC-PUFAS INTAKE AND RISK REDUCTION OF ALLERGIC DISEASES

There are reports demonstrating that fish oil, if consumed during pregnancy and the first year of life, may reduce the risk of Th1 immune response mediated diseases such as early diabetes type I in children. This apparent benefit in Th1 and Th2 mediated diseases supports the idea that changes in environmental factors are immunomodulators, e.g., dietary introduction of ω-3 LC-PUFAS produces its effects by common ways rather than a simple Th1/Th2 deviation effect [78].

Furuhjelm *et al.* reported that ω-3 LC-PUFAS supplementation during pregnancy and nursing in women at high risk to have babies with allergy, may reduce the risks of allergic sensitization to eggs, preventing the development of atopic dermatitis and food allergy associated with IgE production during the first year of life [79].

Recently, a study in Sweden showed that one out of five children suffers from atopic dermatitis, with relevant role of inheritance and with neither protective effects of the nursing period nor the age at which milk and eggs consumption was started to be consumed. However, protective effects were reported when fish was introduced before nine months of age [80].

Moreover, it has been reported that adult women with high fish and DHA intake present low levels of allergic sensitization. The reason why this association has only been found in women is not completely understood. However, it is possible that gender differences regarding PUFAS metabolism could explain it [81].

In conclusion, at this time, there is not enough evidence to recommend ω-3 LC-PUFAS supplementation, neither in the pregnant mother, nor in the breastfeeding mother in order to prevent allergic disease development in high risk population. However, current studies are too heterogeneous and consider only ω-3 LC-PUFAS intake at different doses. We need to develop more controlled studies with high dose supplementation of ω-3 LC-PUFAS (2-4 grams per day with EPA and DHA) to elucidate that issue. Similarly, there are no conclusive data about the prevention of allergy development in high risk infants feeding with ω-3 LC-PUFAS supplementation.

ω-3 LC-PUFAS consumption may reduce the use of anti-inflammatory drugs in the treatment of patients with asthma and EIB, probably because anti-inflammatory drugs and ω-3 LC-PUFAS exert their therapeutic effects, almost in part, through the same molecular actions. There may be a role of lipid mediators derived from ω-3 LC-PUFAS metabolism (lipoxins and resolvins) in the resolution of allergic asthma inflammation. Hence, interactions between food (especially fish oil) and drugs may take place and may show more anti-inflammatory advantages than the simple pharmacological intervention, therefore, the use of PUFAS supplementation could represent an adjunctive therapy in asthma (secondary prevention). Placebo-controlled studies with a high-quality methodology design are required so as to draw better conclusions, especially with the employment of ω-3 LC-PUFAS at high doses.

REFERENCES:

[1]   Bousquet, J.; Khaltaev, N.; Cruz, A. A.; Denburg, J.; Fokkens W. J.; Togias, A.; Zuberbier, T.; Baena-Cagnani, C. E.; Canonica, G. W.; van Weel, C.; Agache, I.; Aït-Khaled, N.; Bachert, C.; Blaiss, M.; Bonini, S.; Boulet, L. P.; Bousquet, P. J.; Camargos, P.; Carlsen, K. H.; Chen, Y.; Custovic, A.; Dahl, R.; Demoly, P.; Douagui, H.; Durham, S. R.; van Wijk, R. G.; Kalayci, O.; Kaliner, M. A.; Kim, Y. Y.; Kowalski, M. L.; Kuna, P.; Le, L. T.; Lemiere, C.; Li, J.; Lockey, R. F.; Mavale-Manuel, S.; Meltzer, E. O.; Mohammad, Y.; Mullol, J.; Naclerio, R.; O'Hehir, R. E.; Ohta, K.; Ouedraogo, S.; Palkonen, S.; Papadopoulos, N.; Passalacqua, G.; Pawankar, R.; Popov, T. A.; Rabe, K. F.; Rosado-Pinto, J.; Scadding, G. K.; Simons, F. E.; Toskala, E.; Valovirta, E.; van Cauwenberge, P.; Wang, D. Y.; Wickman, M.; Yawn, B. P.; Yorgancioglu, A.; Yusuf, O. M.; Zar, H.; Annesi-Maesano, I.; Bateman, E. D.; Ben Kheder, A.; Boakye, D. A.; Bouchard, J.; Burney, P.; Busse, W. W.; Chan-Yeung, M.; Chavannes, N. H.; Chuchalin, A.; Dolen, W. K.; Emuzyte, R.; Grouse, L.; Humbert, M.; Jackson, C.; Johnston, S. L.; Keith, P. K.; Kemp, J. P.; Klossek, J. M.; Larenas-Linnemann, D.; Lipworth, B.; Malo, J. L.; Marshall, G. D.; Naspitz, C.; Nekam, K.; Niggemann, B.; Nizankowska-Mogilnicka, E.; Okamoto, Y.; Orru, M. P.; Potter, P.; Price, D.; Stoloff, S. W.; Vandenplas, O.; Viegi, G.; Williams, D.; World Health Organization; GA(2)LEN; AllerGen. Allergic rhinitis and its impact on asthma (ARIA) 2008 update (in collaboration with the World Health Organization, GA^{2}LEN and AllerGen). *Allergy*, **2008**, *63*, 8-160.

[2]   De Antueno, R.J.; Knickle, L.C.; Smith, H.; Elliot, M.; Allen, S.J.; Nwaka, S.; Winther, M.D. Activity of human D5 and D6 desaturases on multiple n-3 and n-6 polyunsaturated fatty acids. *FEBS Letters*, **2001**, *1*, 77-80.

[3]   Towsend, C. M.; Beauchamp, M. D.; Evers B.M.; Mattox, K. L. *Sabiston Textbook of Surger, 18th ed.*; Sounders Elsevier: Philadelphia, **2007**.

[4]   D'Andrea, S.; Guillou, H.; Jan, S.; Catheline, D.; Thibault, J.; Bouriel, M.; Rioux, V.; Legrand, P. The same rat D6-desaturase not only acts on 18-but also on 24-carbon fatty acids in very-long chain polyunsaturated fatty acid biosynthesis. *Biochem. J.*, **2002**, *364*, 49-55.

[5]   Nelson, D. L.; Cox, M. M. Lehninger Principles of Biochemistry, 4th ed.; Worth Publishers: New York, **2000**.

[6]   Simopoulos, A. P. Omega-3 fatty acids in health and disease and in growth and development. *Am. J. Clin. Nutr.*, **1991**, *54*, 438-63.

[7]   Romagnani, S. The Th2 hypothesis in allergy: Eppur si muove! *Allergy Clin. Immunol. Intern.*, **1998**, *10*, 158–165.

[8]   Rottem, M.; Shoenfeld, Y. Asthma as a paradigm for autoimmune disease. *Int. Arch. Allergy Immunol.*, **2003**, *132*, 210–214.

[9]   Kimura, M.; Yamaide, A.; Tsuruta, S.; Okafuji, I.; Yoshida, T. Development of the capacity of peripheral blood mononuclear cells to produce IL-4, IL-5 and IFN-gamma upon stimulation with house dust mite in children with atopic dermatitis. *Int. Arch. Allergy Immunol.*, **2002**, *127*, 191–197.

[10]   Ng, T. W.; Holt, P.G.; Prescott, S. L. Cellular immune responses to ovalbumin and house dust mite in egg-allergic children. *Allergy*, **2002**, *57*, 207–214.

[11]   Smart, J. M.; Kemp, A. S. Increased Th1 and Th2 allergen-induced cytokine responses in children with atopic disease. *Clin. Exp. Allergy*, **2002**, *32*, 796–802.

[12]   Grassegger, A.; Hopfl, R. Significance of the cytokine interferon gamma in clinical dermatology. *Clin. Exp. Dermatol.*, **2004**, *29*, 584–588.

[13]   Cho, S. H.; Stanciu, L. A.; Begishivili, T.; Bates, P. J.; Holgate, S. T.; Johnston, S. L. Peripheral blood CD4+ and CD8+ T cell type 1 and type 2 cytokine production in atopic asthmatic and normal subjects. *Clin. Exp. Allergy*, *2002*, *32*, 427–433.

[14]   Wills-Karp, M.; Santeliz, J.; Karp, C. L. The germless theory of allergic disease: revisiting the hygiene hypothesis. *Nat. Rev. Immunol.*, *2001*, *1*, 69–75.

[15]   Lange, J.; Ngoumou, G.; Berkenheide, S.; Moseler, M.; Mattes, J.; Kuehr, J.; Kopp, M. V. High interleukin-13 production by phytohaemagglutinin- and Der p 1-stimulated cord blood mononuclear cells is associated with the subsequent development of atopic dermatitis at the age of 3 years. *Clin. Exp. Allergy*, *2003*, *33*, 1537–1543.

[16]   Neaville, W. A.; Tisler, C.; Bhattacharya, A.; Anklam, K.; Gilbertson-White, S.; Hamilton, R.; Adler, K; Dasilva, D. F.; Roberg, K. A.; Carlson-Dakes, K. T.; Anderson, E.; Yoshihara, D.; Gangnon, R.; Mikus, L. D.; Rosenthal, L. A.; Gern, J. E.; Lemanske, R. F. Developmental cytokine response profiles and the clinical and immunologic expression of atopy during the first year of life. *J. Allergy Clin. Immunol.* *2003*, *112*, 740–746.

[17]   Upham, J. W.; Holt, P. G.; Taylor, A.; Thorton, C. A.; Prescott, S.L. HLA-DR expression on neonatal monocytes is associated with allergen-specific immune responses. *J. Allergy Clin. Immunol.*, *2004*, *114*, 1202–1208.

[18]   Hagendorens, M. M.; Ebo, D. G.; Bridts, C. H.; Van de Water, L.; De Clerck, L. S.; Stevens, W. J. Prenatal exposure to house dust mite allergen (Der p 1), cord blood T cell phenotype and cytokine production and atopic dermatitis during the first year of life. *Pediatr. Allergy Immunol.*, *2004*, *15*, 308–315.

[19]   Rowe, J.; Heaton, T.; Kusel, M.; Suriyaarachchi, D.; Serralha, M.; Holt, B. J.; de Klerk, N.; Sly, P. D.; Holt, P. G. High IFN-gamma production by CD8+ T cells and early sensitization among infants at high risk of atopy. *J. Allergy Clin. Immunol.*, *2004*, *113*, 710–716.

[20]   Macaubas, C.; de Klerk, N. H.; Holt, B. J.; Wee, C.; Kendall, G.; Firth, M.; Sly, P. D.; Holt, P.G. Association between antenatal cytokine production and the development of atopy and asthma at age 6 years. *Lancet*, *2003*, *362*, 1192–1197.

[21]   Khair-el-Din, T. A.; Sicher, S. C.; Vázquez, M. A.; Wright, W. J.; Lu, C. Y. Docosahexaenoic acid, a major constituent of fetal serum and fish oil diets, inhibits IFN gamma-induced Ia-expression by murine macrophages in vitro. *J. Immunol.*, *1995*, *154*, 1296–1306.

[22]   Reece, M. S.; McGregor, J. A.; Allen, K. G.; Harris, M. A. Maternal and perinatal long chain fatty acids: possible roles in preterm birth. *Am. J. Obstet. Gynecol.*, *1997*, *176*, 907–914.

[23]   Wegmann, T.G.; Lin, H.; Guilbert, L.; Mosmann, T. R. Bidirectional cytokine interactions in the maternal–fetal relationship: is successful pregnancy a TH2 phenomenon? *Immunol. Today*, *1993*, *14*, 353–356.

[24]   Miles, E.; Aston, L.; Calder, P. C. *In vitro* effects of eicosanoids derived from different 20-carbon fatty acids on T helper type 1 and T helper type 2 cytokine production in human whole-blood cultures. *Clin. Exp. Allergy*, *2003*, *33*, 624–632.

[25]   Jones, A. C.; Miles, E. A.; Warner, J. O.; Colwell, B. M.; Bryant, T. N.; Warner, J. A. Fetal peripheral blood mononuclear cell proliferative responses to mitogenic and allergenic stimuli during gestation. *Pediatr. Allergy Immunol.*, *1996*, *7*, 109–116.

[26]   Hassan, J.; Reen, D. J. Human recent thymic emigrants: identification, expansion, and survival characteristics. *J. Immunol.*, *2001*, *167*, 1970–1976.

[27]   Adkins, B.; Bu, Y.; Guevara, P. The generation of Th memory in neonates versus adults: prolonged primary Th2 effector function and impaired development of Th1 memory effector function in murine neonates. *J. Immunol.*, *2001*, *166*, 918–925.

[28]   Thornton, C. A.; Upham, J. W.; Wikstrom, M. E.; Holt, B. J.; White, G. P.; Sharp, M. J.; Sly, P. D.; Holt, P. G. Functional maturation of CD4+CD25+CTLA4+CD45RA+ T regulatory cells in human neonatal T cell responses to environmental antigens/allergens. *J. Immunol.*, *2004*, *173*, 3084–3092.

[29]   Galli, E.; Picardo, M.; Chini, L.; Passi, S.; Moschese, V.; Terminali, O.; Paone, F.; Fraioli, G.; Rossi, P. Analysis of polyunsaturated fatty acids in newborn sera: a screening tool for atopic disease? *Br. J. Dermatol.*, *1994*, *130*, 752–756.

[30]   Yu, G.; Kjellman, N. I. Bjorksten, B. Phospholipid fatty acids in cord blood: family history and development of allergy. *Acta Paediatr.*, *1996*, *85*, 679–683.

[31]   Beck, M.; Zelczak, G.; Lentze, M. J. Abnormal fatty acid composition in umbilical cord blood of infants at high risk of atopic disease. *Acta Paediatr.*, *2000*, *89*, 279–284.

[32]   Newson, R. B.; Shaheen, S. O.; Henderson, A. J.; Emmett, P. M.; Sherriff, A.; Calder, P. C. Umbilical cord and maternal blood red cell fatty acids and early childhood wheezing and eczema. *J. Allergy Clin. Immunol.*, *2004*, *114*, 531–537.

[33]   Duchen, K.; Yu, G.; Bjorksten, B. Atopic sensitization during the first year of life in relation to long chain polyunsaturated fatty acid levels in human milk. *Pediatr. Res.*, *1998*, *44*, 478–484.

[34]   Stoney, R. M.; Woods, R. K.; Hosking, C. S.; Hill, D. J.; Abramson, M. J.; Thien, F. C. Maternal breast milk long-chain n-3 fatty acids are associated with increased risk of atopy in breastfed infants. *Clin. Exp. Allergy*, *2004*, *34*, 194–200.

[35]   Calder, P.C. N-3 polyunsaturated fatty acids and inflammation: from molecular biology to the clinic. *Lipids*, *2003*, *38*, 343–352.

[36] Yaqoob, P. Lipids and the immune response: from molecular mechanisms to clinical applications. Curr. *Opin. Clin. Nutr. Metab. Care*, **2003**, *6*, 133–150.

[37] Field, C. J.; Schley, P. D. Evidence for potential mechanisms for the effect of conjugated linoleic acid on tumor metabolism and immune function: lessons from n-3 fatty acids. *Am. J. Clin. Nutr.*, **2004**, *79*, 1190S–1198S.

[38] Mori, T. A.; Beilin, L. J. Omega-3 fatty acids and inflammation. *Curr. Atheroscler. Rep.*, **2004**, *6*, 461–467.

[39] Spector, S. L.; Surette, M. E. Diet and asthma: has the role of dietary lipids been overlooked in the management of asthma? *Ann. Allergy Asthma Immunol.*, **2003**, *90*, 371–377.

[40] Surette, M. E.; Koumenis, I. L.; Edens, M. B.; Tramposch, K. M.; Clayton, B.; Bowton, D.; Chilton, F. H. Inhibition of leukotriene biosynthesis by a novel dietary fatty acid formulation in patients with atopic asthma: a randomized, placebo-controlled, parallel-group, prospective trial. *Clin. Ther.*, **2003**, *25*, 972–979.

[41] Serhan, C. N.; Gotlinger, K.; Hong, S.; Arita, M. Resolvins, docosatrienes, and neuroprotectins, novel omega-3-derived mediators, and their aspirin triggered endogenous epimers: an overview of their protective roles in catabasis. *Prostaglandins & Other Lipid Mediat.*, **2004**, *73*, 155–172.

[42] Ma, D. W.; Seo, J.; Switzer, K. C.; Fan, Y. Y.; McMurray, D. N.; Lupton, J. R.; Chapkin, R. S. n-3 PUFA and membrane microdomains: a new frontier in bioactive lipid research. *J. Nutr. Biochem.*, **2004**, *15*, 700–706.

[43] Fan, Y. Y.; McMurray, D. N.; Ly, L. H.; Chapkin, R. S. Dietary (n-3) polyunsaturated fatty acids remodel mouse T-cell lipid rafts. J. *Nutr.*, **2003**, *133*, 1913–1920.

[44] Switzer, K. C.; Fan, Y. Y.; Wang, N.; McMurray, D. N.; Chapkin, R. S. Dietary n-3 polyunsaturated fatty acids promote activation-induced cell death in Th1-polarized murine CD4þ T-cells. *J. Lipid Res.*, **2004**, *45*, 1482–1492.

[45] Lee, J. Y.; Plakidas, A.; Lee, W. H.; Heikkinen, A.; Chanmugam, P.; Bray, G.; Hwang, D. H. Differential modulation of Toll-like receptors by fatty acids: preferential inhibition by n-3 polyunsaturated fatty acids. *J. Lipid. Res.*, **2003**, *44*, 479–486.

[46] Lapillonne, A.; Clarke, S.; Heird, W. Polyunsaturated fatty acids and gene expression. *Curr. Opin. Clin. Nutr. Metab. Care*, **2004**, *7*, 151–156.

[47] Sampath, H.; Ntambi, J.M. Polyunsaturated fatty acid regulation of gene expression. *Nutr. Rev.*, **2004**, *62*, 333–339.

[48] Simopoulus, A. P. Omega-3 fatty acids in inflammation and autoimmune diseases. *Journal of the American College of Nut.*, **2002**, 6, 495-505.

[49] Kremer, J. M. n-3 Fatty acid supplements in rheumatoid arthritis. *Am. J. Clin. Nutr.*, **2000**, *71*, 349S-351S.

[50] Rodríguez, C. M.; Tovar, A. R.; del Prado, M.; Torres, N. Mecanismos moleculares de acción de los ácidos grasos poliinsaturados y sus beneficios en la salud. *Rev. Invest. Clin.*, **2005**, *57*, 457-472.

[51] Black, P. N.; Sharp, S. Dietary fat and asthma: is there a connection? *Eur. Respir. J.*, **1997**, *10*, 6-12.

[52] Woods, R. K.; Thien, F. C.; Abramson, M. J. Dietary marine fatty acids (fish oil) for asthma in adults and children. *Cochrane Database Syst. Rev.*, *CD001283*, **2002**.

[53] Van Gool CJ, Zeegers MP, Thijs C. Oral essential fatty acid supplementation in atopic dermatitis: a meta-analysis of placebo-controlled trials. *Br. J. Dermatol.*, **2004**, *150*, 728–740.

[54] Oddy, W. H.; de Klerk, N. H.; Kendall, G. E.; Mihrshahi, S.; Peat, J. K. Ratio of omega-6 to omega-3 fatty acids and childhood asthma. *J. Asthma*, **2004**, *41*, 319–326.

[55] Peat, J. K.; Mihrshahi, S.; Kemp, A. S.; Marks, G. B.; Tovey, E. R.; Webb, K.; Mellis, C.M.; Leeder, S. R. Three-year outcomes of dietary fatty acid modification and house dust mite reduction in the Childhood Athma Prevention Study. *J. Allergy Clin. Immunol.*, **2004**, *114*, 807-13.

[56] Anandan, C.; Nurmatov, U.; Sheikh, A. Omega 3 and 6 oils for primary prevention of allergic disease: systematic review and meta-analysis. *Allergy*, **2009**, *64*, 840-848.

[57] Mickleborough, T. D. Dietary Omega-3 Polyunsaturated Fatty Acid Supplementation and Airway Hyperresponsiveness in Asthma. *Journal of Asthma*, **2005**, *42*, 305–314.

[58] Lai, C. K.; Beasley, R.; Crane, J.; Foliaki, S.; Shah, J.; Weiland, S.; International Study of Asthma and Allergies in Childhood Phase Three Study Group. Global variation in the prevalence and severity of asthma symptoms: phase III three of the International Study of Asthma in Childhood Phase Study Group (ISAAC). *Thorax*, **2009**, *64*, 476-483.

[59] Thien F. C. K.; Woods, R.; De Luca, S.; Abramson, M. J. Ácidos grasos marinos alimentarios (aceite de pescado) para el asma en adultos y niños (Spanish Translation Cochrane review). *Cochrane Plus Library*, **2008**, *no.2*, available at: http://www.update-software.com.

[60] Reisman, J.; Schachter, H. M.; Dales, R. E.; Tran, K.; Kourad, K.; Barnes, D.; Sampson, M.; Morrison, A.; Gaboury, I.; Blackman, J. Treating asthma with omega-3 fatty acids: where is the evidence? A systematic review. *Complementary and Alternative Medicine*, **2006**, *6*, 1-8.

[61] Bays, H. E.; Tighe, A.P.; Sadovsky, R.; Davidson, M. H. Prescription omega- 3 fatty acids and their lipid effects: physiologic mechanisms of action and clinical implications. *Expert. Rev. Cardiovasc. Ther.*, **2008**, *6*, 391-409.

[62] Stalenhoef, A. F.; de Graaf, J.; Wittekoek, M. E.; Bredie, S. J.; Demacker, P. N.; Kastelein, J. J. Efficacy of concentrated n-3 fatty acids versus gemfibrozil on plasma lipoproteins, low density lipoprotein heterogeneity and oxidability in patients with hypertriglyceridemia. *Atherosclerosis*, **2000**, *153*, 129-138

[63] Troy, C.; Levy, B. D. Chemical mediators and the resolution of airway inflammation. *Allergol. International*, **2008**, *57*, 299-305.

[64] Haworth, O.; Cernadas, M.; Yang, R.; Serhan, C. N.; Levy, B. D. Resolvin E1 regulates interleukin-23, interferon-gamma and lipoxin A to promote resolution of allergic airway inflammation. *Nature Immunol.*, **2008**, *9*, 873-879.

[65] Hisada, T.; Ishisuka, T.; Aoiki, H.; Mori, M. Resolvin E1 as a novel agent for the treatment of asthma. *Expert. Opin. Ther. Targets*, **2009**, *13*, 513-522.

[66] Spector, S. L.; Surette, M. E. Diet and asthma: has the role of dietary lipids been overlooked in the management of asthma? *Ann. Allergy Asthma Immunol.*, **2003**, *90*, 371-377.

[67] Mickleborough, T. D.; Lindley, M. R.; Ionescu, A. A.; Fly, A. D. Protective effect of fish oil supplementation on exercise-induced bronchoconstriction in asthma. *Chest*, 2006, *129*, 39-49.

[68] Mickleborough, T. D.; Rundell, K. W. Dietary polyunsaturated fatty acids in asthma and exercise-induced bronchoconstriction. *Eur. Respir. Journal* **2005**, *59*, 1335-1346.

[69] Bachert, C.; van Cauwenberge, P. The World Health Organization ARIA (allergic rhinitis and its impact on asthma) Initiative. *Chem. Immunol. Allergy*, **2003**, *82*, 119-126.

[70] Ait-Khaled, N.; Pearce, N. H.; Anderson, R.; Ellwood, P.; Montefort, S.; Shah, J.; and the ISAAC Phase Three Study Group. Global map of the prevalence of symptoms of rhinoconjunctivitis in children: The International Study of Asthma and Allergies in Childhood (ISAAC) Phase Three. *Allergy*, **2009**, *64*, 123–148.

[71] Nafstad, P.; Nystad, W.; Magnus, P.; Jaakkola, J. J. K. Asthma and allergic rhinitis at 4 years of age in relation to fish consumption in infancy. *J. Asthma*, **2003**, *40*, 343-348.

[72] Hoff, S.; Seiler, H.; Heinrich, J.; Nieters, A.; Becker, C.; Nagel, G.; Gedrich, K.; Karg, G.; Wolfram, G.; Linseisen, J. Allergic sensitization and allergic rhinitis are associated with n-3 polyunsaturated fatty acids in the diet and red blood cell membranes. *Eur. J. Clin. Nutr.*, **2005**, *59*, 1021-1080.

[73] Miyake, Y.; Sasaki, S.; Tanaka, K.; Ohya, Y.; Miyamoto, S.; Matsunaga, I.; Yoshida, T.; Hirota, Y.; Oda, H.; Osaka Maternal and Child Health Study Group. Fish and Fat Intake and Prevalence of Allergic Rhinitis in Japanese Females: the Osaka Maternal and Child Health Study. *Am. J. Clin. Nutr.*, **2007**, *26*, 279-287.

[74] Asher, M. I.; Montefort, S.; Björkstén, B.; Lai, C. K. W.; Strachan, D. P.; Weiland, S. K.; Williams, H.; and the ISAAC Phase Three Study Group. Worldwide time trends in the prevalence of symptoms of asthma, allergic rhinoconjunctivitis, and eczema in childhood: ISAAC Phases One and Three repeat multicountry cross-sectional surveys. *Lancet*, **2006**, *368*, 733–43.

[75] Bracco, U.; Deckelbaum, R. J. *Polyunsaturated fatty acids in human nutrition*; Raven Press: New York, **1992.**

[76] Horrobin, D. F. Essential fatty acid metabolism and its modification in atopic eczema. *Am. J. Clin. Nutr.*, **2000**, *71*, 367S-372S.

[77] Bieber, T. Atopic Dermatitis. *New England J. Med.*, **2008**, *358*, 1483-94.

[78] Dunstan, J. A.; Prescott, S. L. Does fish oil supplementation in pregnancy reduce the risk of allergic disease in infants? *Curr. Opin. Allergy Clin. Immunol.*, **2005**, *5*, 215-221.

[79] Furuhjelm, C.; Warstedt, K.; Larsson, J.; Fredriksson, M.; Böttcher, M. F.; Fälth-Magnusson, K.; Duchén, K. Fish oil supplementation in pregnancy and lactation may decrease the risk of infant allergy. *Acta Paediatrica*, **2009**, *98*, 1461-7.

[80] Alm, B.; Aberg, N.; Erdes, L.; Molloborg, P.; Pettersson, R.; Norvenius, S. G.; Goksör, E.; Wennergren, G. Early introduction of fish decrease the risk of eczema in infants. *Arch. Dis. Child.*, **2009**, *94*, 11-15.

[81] Schnappinger, M.; Sausenthaler, S.; Linseisen, J.; Hauner, H.; Heinrich, J. Fish Consumption, Allergic Sensitisation and Allergic Diseases in Adults. *Annals of Nutrition and Metabolism*, **2009**, *54*, 667-8.

CHAPTER 8

Recommendations and Sources of n-3 Long-Chain Polyunsaturated Fatty Acids

Mariela Bernabe-Garcia*, M. Sc. and Mardia López-Alarcón, Ph.D

Unidad de Investigación Médica en Nutrición, Hospital de Pediatría, C.M.N. Siglo XXI. Instituto Mexicano del Seguro Social.

Abstract: The n-3 long-chain polyunsaturated fatty acids (LC-PUFAs) eicosapentaenoic acid (EPA) and docosahexaenoic acid (DHA) have demonstrated a beneficial effect on reducing morbidity and mortality of some highly prevalent chronic diseases. Therefore, there is considerable interest in establishing recommendations for EPA and DHA. The Institute of Medicine of The National Academies in 2002 established that 10% (~100 mg/day) of the acceptable macronutrient distribution range from α-linolenic acid (ALA) can be provided from EPA and DHA in order to avoid deficiencies and to support adequate neurodevelopment and growth. However, most of these data were obtained from epidemiological investigations rather than from clinical research. Additionally, this amount represents the average intake of these n-3 fatty acids in a healthy population but is not a dietary reference intake (DRI); hence, this amount has been qualified as low according to the scientific community. Considering the enormous health benefits based on several specific effects of n-3 LC-PUFAs, DRIs should be re-evaluated in light of the new evidence and the recommendations of numerous international federal agencies. At the present time, there is evidence of beneficial for prevention of coronary heart disease (CHD) and cardiac death at intakes of 250 to 500 mg/day of EPA + DHA as well for pregnant, lactating and childbearing woman whose daily consumption should be at least 200 mg of DHA. Meanwhile, evidence is inconclusive for preterm infants and other pathological entities such as cognitive decline and affective disorders.

INTRODUCTION

Alpha-linolenic acid (ALA) and linoleic acid (LA) are considered as essential nutrients because mammalian cells are unable to synthesize these, as explained by the absence of Δ-12 and Δ-15 desaturase. Therefore, these nutrients must be obtained from external sources such as diet or dietary supplements. All polyunsaturated fatty acids are characterized by at least a double bond at the third carbon for the n-3 family or sixth carbon for the n-6 family, starting from the methyl end of the fatty acid. ALA and LA are the precursors of the n-3 and n-6 family, respectively, and are converted by chain elongation, desaturation through Δ-6 and Δ-5 desaturase and chain-shortening into their respective long-chain metabolites, collectively named long-chain polyunsaturated fatty acids (LC- PUFAs) and characterized by containing ≥20 carbon atoms and ≥3 double bonds. The most important LC-PUFA derived from LA is arachidonic acid (AA) and it is the most abundant in tissue mammalian reserves. Meanwhile, the most important LC-PUFAs from ALA are eicosapentaenoic acid (EPA, 20:5 n-3), docosapentaenoic acid (DPA, 22:5 n-3), and docosahexaenoic acid (DHA, 22:6 n-3) [1-3] (Fig. **1**).

Fatty acid deficiency in sick or malnourished pediatric or geriatric populations was the main focus for the medical community. However, recent studies about these fatty acids increased the scientific and medical interest when beneficial effects on brain and retina function were observed, as well as their role in minimizing the risk of adverse health outcomes such as cardio-metabolic diseases. Therefore, interest concerning requirements and dietary recommendations has also recently increased [1].

This chapter will review the current evidence and suggested dietary recommendations for n-3 LC-PUFAs during certain physiological stages and chronic diseases.

DIETARY RECOMMENDATIONS FOR ALA, DHA AND EPA

In 2002, the U.S. Dietary Reference Intakes (DRIs) concluded that insufficient data were available to establish n-3 LC-PUFAs requirements or Recommended Dietary Allowances (RDA) for EPA and DHA. Although there is no

*Address correspondence to M. Sc. Mariela Bernabe Garcia: Unidad de Investigación Médica en Nutrición, Apartado Postal C-029 "Coahuila", Coahuila #5, Col. Roma, México, D.F., 06703, México. Tel/Fax (52) 55 5627 6944; E-mail: mariela_bernabe@yahoo.com

DRI for EPA and DHA, an adequate intake (AI) was instead reported as derived from dietary surveys in the U.S. by life stage and gender [1]. The current AI for ALA issued by the Institute of Medicine of the National Academies is 1.6 g/day for men and 1.1 g/day for women between 19 and <70 years of age [4] and corresponds to an acceptable macronutrient distribution range (AMDR) of 0.6–1.2% of the total energy intake per day [5]. However, because those data were based on descriptors of U.S. population intakes, it is important to recognize that they are not based on health or biochemical endpoints and are, therefore, not sufficiently supported [1].

Accordingly, ~10% of the AMDR for ALA can be consumed as EPA and DHA. The DRI report for macronutrients stated that scientific reports suggest that higher intakes of ALA, EPA and DHA may confer some degree of protection to prevent coronary heart disease (CHD). However, because the physiological potency of EPA and DHA is greater than ALA, it was not possible to estimate an AMDR for each n-3 fatty acid [5].

In 2008, the Technical Committee on Dietary Lipids of the International Life Sciences Institute (ILSI) of North America re-examined the new evidence supporting the need to reassess this recommendation considering whether the body of evidence specific to the major chronic diseases in the U.S. such as CHD and cognitive decline, among others, has evolved sufficiently to justify the reconsideration of DRIs for EPA and DHA [6].

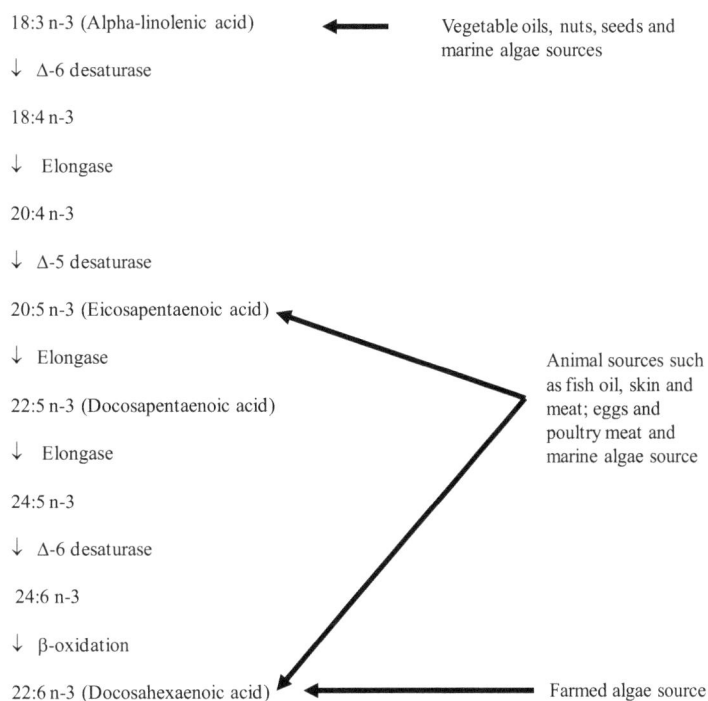

18:3 n-3 (Alpha-linolenic acid) ⟵ Vegetable oils, nuts, seeds and marine algae sources

↓ Δ-6 desaturase

18:4 n-3

↓ Elongase

20:4 n-3

↓ Δ-5 desaturase

20:5 n-3 (Eicosapentaenoic acid)

↓ Elongase

22:5 n-3 (Docosapentaenoic acid)

↓ Elongase

24:5 n-3

↓ Δ-6 desaturase

24:6 n-3

↓ β-oxidation

22:6 n-3 (Docosahexaenoic acid) ⟵ Farmed algae source

Animal sources such as fish oil, skin and meat; eggs and poultry meat and marine algae source

Figure 1: Metabolic pathway of elongation and desaturation to synthesize long-chain polyunsaturated fatty acids from their precursors and main dietary sources. Modified from Innis SM [1].

Cardiovascular Disease

A recent ILSI report described that CHD has a spectrum of biological processes, related risk factors and mechanical pathways, each of which may be affected by particular risk factors and interventions including n-3 LC-PUFAs. The most important effect of n-3 LC-PUFAs consumption is to reduce cardiac death, which was based on consistent and strong evidence from prospective cohort studies in healthy population, one case-control study of sudden cardiac death (SCD) and four large randomized controlled trials with fish or fish oil in patients with and without known disease. The pooled analysis of observational studies for primary prevention (subjects without CHD) and from controlled trials in subjects for primary and secondary prevention (with or without CHD) demonstrated that a modest consumption of ~250–500 mg/day of EPA + DHA lowers the relative risk by 25% or more compared with little or no intake, although in a nonlinear pattern. According to the authors, higher intakes do not substantially further lower CHD mortality, suggesting a threshold effect. At intakes up to 250 mg/day of EPA + DHA, the relative risk of CHD death was 14.6% lower [95% confidence interval (CI) from 8% to 21%] per each 100 mg/day of EPA + DHA, for a total risk reduction of 36% (95% CI, 20–50%) [7]. However, the effect of EPA + DHA in generally healthy individuals was considered to be most relevant for establishing population DRIs for other authors [6]. Thus, a meta-analysis of estimated dietary EPA + DHA consumption was carried out including 4473 cardiac deaths in 326,572 generally healthy individuals in 16 prospective cohort studies from Europe, U.S., China and Japan, concluding that an intake of at least 250 mg of EPA + DHA reduces the risk for CHD [6]. Other recent meta-analyses, which included six U.S. prospective studies, established that intakes of ~500 mg/day conferred the highest protection for preventing CHD death [8]. One study in males with angina reported no benefit of dietary advice of the increased fish consumption [9]. More modest magnitudes of beneficial effects have been reported for atherosclerosis progression and nonfatal acute coronary syndrome [6,7]. Heterogeneity of fish or fish oil intake on intermediate cardiovascular outcomes is likely related to varying dose and time responses of effects on the risk factors related to CHD [7].

The relationship of dietary EPA + DHA and all-cause mortality or total mortality has also been analyzed. n-3 PUFAs is considered to be the strongest factor affecting death due to CHD (but no other causes of mortality). Therefore, total mortality in a population will depend on the proportion (prevalence) of deaths due to CHD. As an example, prevalence of CHD death has been reported to be 25% in middle-aged populations [10], reaching 50% in populations with diagnosed CHD [11]. A reduction of ~36% in CHD attributable to dietary EPA + DHA would be expected to result in a reduction of total mortality between ~9% and 18% (on average of ~14% in a mixed population) [7]. When two additional randomized double-blind studies with subjects at high risk of arrhythmias were added [12,13], marine n-3 PUFAs reduced up to 17% (pooled relative risk of 0.83; 95% CI 0.68 to 1.00; p = 0.046 [7]. In conclusion, there is strong evidence that consumption of two portions of fish or one portion of oily fish containing between 250 to 500 mg/day would be useful for prevention of primary or secondary CHD.

Worldwide Recommendations for EPA DHA, and ALA for Primary Prevention of Coronary Diseases

Because health benefits are associated with EPA and DHA, numerous agencies and organizations worldwide have issued recommendations for fish, fish oil, and EPA + DHA, in addition to ALA intake for healthy adults and for secondary prevention of CHD (Table 2). For healthy adults, ALA recommendations vary from 1.6 to 2 g/day (0.6–2% of total energy intake) [14-17]. Meanwhile, for EPA + DHA, dietary recommendations range from one to three servings per week of fish (preferably oily fish) or from 250 to 500 mg/day of a combination of EPA and DHA [14,18-20,22-27], except for Australian men in whom the dietary target is 610 mg/day of n-3 LC-PUFAs [21]. For secondary CHD prevention, the amount recommended is, as expected, higher than for healthy adults: from two servings per week or an almost daily serving of fatty fish (preferably from oily fish) or EPA + DHA supplements within a balanced diet to provide up to 250 mg/day until reaching ~1 g/day of EPA + DHA [19,25-27]. For hypertriglyceridemia >500 mg/dL, recommendations are from 2 to 4 g/day of EPA and DHA provided as capsules under a physician's care as part of the treatment with a low fat diet, physical activity, etc., as established by the American Heart Association [25].

Table 1: Recommendations of n-3 fatty acids or fish consumption for primary and secondary CHD prevention by international organizations

Organization or Council and year	Recommendations
American Heart Association, 2002 [25]	Healthy adults: 2 servings/week of fish including oils and foods rich in ALA. CHD patients: ~1 g of EPA + DHA/day, both preferably from fatty fish. Hypertriglyceridemia >500 mg/dL: 2–4 g/day of EPA + DHA.
European Society for Cardiology, 2003 [22]	Oily fish and n-3 fatty acids have particular protective properties for prevention of primary cardiovascular disease.
Food and Agriculture Organization/WHO, 2003 [14]	1–2 servings/week of fish, each one should provide 200–500 mg of EPA + DHA. Total n-3 fatty acids, 1%–2% of energy intake. Vegans consume an adequate consumption of ALA (1–2%).
UK Scientific Advisory Committee on Nutrition, 2004 [18]	At least 2 portions of fish/week: one should be oily. Two portions of fish per week (one white and one oily contain aproximately 450 mg/day of n-3 long chain polyunsaturated fatty acids.
Superior Health Council of Belgium, 2004 [19]	Healthy adults: 2 servings/week, min. 0.3% of energy from EPA + DHA (~667 mg/day). For ALA, use rapeseed, soybean oils or blends of n-3 and monounsaturated fatty acids as olive oil. For secondary prevention, an almost daily serving of fatty fish or, alternatively, 1 g/day of EPA+DHA in fish oil capsules within a balanced diet.

	Table 1: cont....
International Society for Study of Fatty Acids and Lipids, 2004 [16]	500 mg/day minimum of a combination of EPA+DHA. For ALA intake, ~0.7% of energy.
The Dietary Guidelines for Americans, 2005 [26]	Evidence suggests consuming ~2 servings of fish/week (~8 oz or 227 g) may reduce the risk of CHD mortality, and consuming EPA and DHA may reduce the risk of cardiovascular mortality after a cardiac event.
National Academies, Institute of Medicine, 2005-2006 [4,17]	Men and women 19 to >70 years, at least 1.6 and 1.1 g/day of ALA, respectively, or 0.6%-1.2% of ALA from energy intake. Seafood is part of a healthy diet and can be substituted for other protein sources higher in saturated fat.
Health Council of the Netherlands, 2006 [20]	Eat fish twice weekly, one of which should be oily fish to achieve the dietary intake of 450 mg/day of n-3 fatty acids from fish.
Australia and New Zealand National Health and Medical Research Council, 2006 [21]	For men and women 19 to >70 years, at least 160 mg/day and 90 mg/day of n-3 LC-PUFAs, respectively, and a dietary target of 610 and 430 mg/day of DHA/EPA/docosapentaenoic acid, respectively.
American Dietetic Association /Dietitians of Canada, 2007 [23]	500 mg/day of EPA + DHA provided by two servings [one serving is 4 oz (112 g) cooked) of fatty fish/week].
American Diabetes Association, 2008 [24]	Two or more servings/week of fish (without considering commercially fried fish filets) to provide n-3 LC-PUFAs
National Heart Foundation of Australia, 2008 [15]	Consume ~500 mg/day of DHA and EPA through a combination of two to three servings (150 g) of oily fish/week, fish oil or food and drink enriched with marine n-3 PUFAs. For ALA, consume at least 2 g/day.

CHD: coronary heart disease; EPA: eicosapentaenoic acid; DHA: docosahexaenoic acid; ALA: alpha-linolenic acid.

A very high n-6:n-3 ratio of fatty acids has been considered an important determinant in the resulting highly prevalent inflammatory-based conditions including cardiovascular, inflammatory and autoimmune diseases, as well as a higher mortality for cardiovascular disease. The ratios of dietary n-6:n-3 fatty acids estimated from the late Paleolithic Age showed a high consumption of n-3 fatty acids compared with current Western diets (U.S.). LA:ALA ratio previously was 0.70 compared with 18.75, respectively. AA + docosatetranoic acid:EPA + DPA + DHA previously was 1.79 vs. 3.33, respectively. Total n-6:n-3 ratio previously was 0.79 (~1:1) compared to 16.74 (range: 15–20:1), respectively. Therefore, it is advisable to reach closer to a ratio of 1:1 as in the case of Greenland Eskimos where mortality for Cardiovascular Disease is lowest [28].

Cognitive Decline with Age

Alzheimer's disease (AD) is the most common form of dementia. Several studies have suggested that cardiovascular risk factors are also involved. Additionally, it has been suggested that the relationship between diet and AD is similar to that between diet and coronary disease, involving antioxidants, fish, dietary fats, and B-vitamins, among others. Pleiotropic effects are exerted by n-3 fatty acids on cardiovascular and central nervous system and may be protective against age-related cognitive decline [29,30]. Moreover, reports of high blood levels of n-6 compared to n-3 fatty acids were associated with AD and increased cognitive decline [31-33]. In seven epidemiological studies, it has been reported that increase in either fish or n-3 fatty acids intake was associated with reduced risk for cognitive decline or dementia in between 40 and 50% of subjects [34-40]. There was no beneficial effect reported in two other studies when controlling by the presence of the gene confounder ApoE4 [41,42]. However, a recent epidemiological study did not find beneficial effects of dietary consumption of fish and n-3 PUFAs in 5395 participants in relation to long-term risk of dementia after a follow-up of 9.6 years [43]. In a randomized trial, 175 subjects received a daily intake of 1.7 g of DHA and 0.6 g of EPA or placebo during 6 months. In a subgroup of 32 subjects with very mild cognitive dysfunction, a significant reduction was observed in rate of decline in the n-3 fatty acid-treated group compared with the placebo group [44]. To date, differences in the severity of dementia, dosages of DHA and EPA, duration and instruments to measure the outcomes still remain inconclusive. A preliminary recommendation advises two to three servings per week of fish to lower risk of AD and cognitive decline with aging [6].

Affective Disorders

There is some evidence that low levels of n-3 PUFAs in tissues are associated with depression and bipolar disorder [45]. Because blood levels of polyunsaturated fatty acids are accepted biomarkers of status [46], biological plausibility is suggested. A recent review stated that the dietary intake required for prevention or treatment of affective disorders should maintain 7% of erythrocyte DHA composition to be an acceptable target. Daily doses required to achieve this may be informative [47]. According to the author, in U.S. subjects, administration of 200 mg/day of DHA or 104 mg/day of EPA + 378 mg/day of DHA for 12 months and 16 weeks, respectively, did not reach the proposal of 7% of erythrocyte DHA composition [48, 49]. Higher doses such as 1296 mg/day of EPA + 864 mg/day of DHA or 1 g of DHA increased erythrocyte EPA + DHA composition to ~8% [48,50]. Taking these studies into consideration, a daily dose has been suggested at between 700 and 1000 mg/day for adults, leading to a stable 7 to 8% in erythrocyte DHA composition [47]. Another recent review reported that three out of five interventional studies showed no effect of n-3 fatty acids supplementation on depressive symptoms. These studies administered supplements of 630 mg/day of EPA + 850 mg/day of DHA (for 12 weeks) until reaching 2.2 g/day of DHA + 0.6 g/day of EPA (during a period of 16 weeks). Meanwhile, the two studies that showed improved ratings of depression administered 2.2 g/day of EPA and 1.2 g/day of DHA until reaching 4.4 g/day of EPA and 2.2 g/day of DHA (for 8 weeks). Although one determinant of the effectiveness of n-3 LC-PUFAs supplementation may be a low initial blood status [45], LC-PUFAs doses clearly overlap among beneficial and no-effect studies. Thus, evidence is still conflicting. Nevertheless, other authors suggested the following recommendations: for postpartum depression and bipolar depression, 0.15% of energy from EPA + DHA showed major benefits; n-3 fatty acids may improve psychotic, aggressive and depressive symptoms in severe patients, whereas for major depression an intake of 0.35% to 0.40% from energy such as EPA + DHA showed marked benefits [51]. More studies in high-risk populations with longer follow-up are needed.

Recommendations for LC-PUFAs in Pregnancy, Lactation and Infancy

It is well known that LC-PUFAs are important for brain and retina functions as well as for infant growth and development [1]. Moreover, LC-PUFAs have the potential for long-lasting effects that extend beyond the period of dietary insufficiency [52]. Although it has been reported that ALA is converted to EPA more efficiently in young non-pregnant woman than in men [53], in most cases these studies have shown little or no effects on DHA content on plasma fractions, circulating blood cells or breast milk [46]. It is also known that human milk is relatively low in DHA. The mean concentration by weight is 0.32 ± 0.22 (range: 0.06–1.4%) from healthy mothers of term infants who consumed free or controlled diets during interventional studies [54]. Therefore, this DHA concentration is definitely insufficient for very-low-birth-weight infants who have an intrauterine accretion of n-3 fatty acids ~50 mg/kg/day [55] and insufficient capacity for elongation and desaturation to achieve their requirements [52]. Smithers *et al.* proposed that the high requirement in preterm infants is perhaps not totally covered with current DHA-supplemented formulas [56]. These authors found that infants fed milk with DHA concentrations at 1% of total fatty acids had improved visual acuity at 4 months of corrected age over infants receiving standard DHA doses of ~0.3%. Therefore, for premature infants, especially those with birth weight <1250 g, increases of the DHA content in human milk to ~1–1.5% of fatty acids has been suggested, either by providing mothers with a DHA supplement or by adding DHA directly to the milk [52,56]. Pregnant and lactating women who attempt to meet the DHA recommendation from fish consumption are advised to be cautious about certain fish because these may be sources of contaminants such as dioxins, polychlorinated biphenyls (PCBs), brominated flame retardants, camphechlor, organotin and methylmercury. Methylmercury has the highest potential for toxicity. Concentrations of methylmercury in aquatic species depend on levels of environmental contamination, lifespan of the species and their predatory nature. U.S. fish with the highest methylmercury content are shark, swordfish, pike, marlin, mackerel (king), and golden bass from the Gulf of Mexico (tilefish), whereas smaller or shorter-lived species (e.g., shellfish, farmed or wild salmon, shrimp, scallops) have lower concentrations. This is relevant because during early development, methylmercury is particularly toxic for the developing brain and may also adversely affect growth in children [5,7,57]

An expert review of the current knowledge [58] supported by the World Association of Perinatal Medicine, the Early Nutrition Academy, and the Child Health Foundation established the following recommendations:

> ➤ Pregnant and lactating women should aim at achieving an average DHA intake of at least 200 mg/day, although intakes up to 1 g/day of DHA or 2.7 g/day of n-3 LC-PUFAs have been used in randomized clinical trials without significant adverse effects.

> ➤ Women of childbearing age can meet the recommended intake of DHA by consuming one or two portions of sea fish per week, including fatty fish (similar to primary prevention of CHD in healthy subjects). However, dietary fish should be selected from a wide range of species without undue preference for predator fish, which are more likely to be contaminated with methylmercury.

> ➤ Intake of the precursor, ALA, is far less effective with respect to DHA deposition in fetal brain than the intake of preformed DHA.

> ➤ There is no evidence that women of childbearing age, whose dietary intake of linoleic acid is adequate, need additional dietary intake of arachidonic acid.

> ➤ Screening for dietary inadequacies should be performed during pregnancy, preferably during the first trimester. If less than desirable dietary habits are detected, individual counseling should be offered during pregnancy as well as during lactation.

In summary, pregnant and lactating women, as well as women of childbearing age, should consume one to two portions of fish per week from nonpredator species in order to achieve at least an intake of 200 mg/day of DHA. ALA is not effective as a food source due to its very low conversion to DHA.

Recommendations for LC-PUFAs for Infants

Koletzko B., *et al* (2008), [58] strongly endorse breastfeeding as the preferred method of feeding healthy infants. They also emphasize, however, the importance of the provision of a balanced dietary intake for breastfeeding women, and this should include a regular supply of DHA. When breastfeeding is not possible, available evidence supports the addition of DHA to infant formula. The addition of at least 0.2% of fatty acids as DHA in formulas appears necessary for achieving a benefit on functional endpoints, but DHA levels should not exceed 0.5% of fatty acids because systematic evaluation of higher levels of intake have not been published. Infant formula contents of AA should be at least those of added DHA, and EPA should not exceed levels of DHA (see Table **2** for several formulas available in the market). In addition, dietary LC-PUFAs supply should continue during the second 6 months of life, but there is insufficient information for quantitative recommendations for addition of LC-PUFAs to follow-on formula or complementary foods. This review [58] followed the present guidelines from an ESPGHAN Coordinated International Expert Group providing recommendations on the composition of infant formula, the Codex Alimentarius proposal for a global infant formula standard and the EU Commission Directive on infant formula and follow-on formula [59-61]. This recommendation guideline was also adopted by the International Society for the Study of Fatty Acids and Lipids [62].

Table 2: Energy, lipids and n-3 fatty acid content in infant formula.

Infant formula*	Content in 100 g of portion			
	Energy, kcal	Lipids, g	AA, g	DHA, g
ABBOTT NUTRITION, SIMILAC NATURAL CARE, ADVANCE, ready-to-feed, ARA & DHA[a]	78	4.24	0.018	0.11
ABBOTT NUTRITION, SIMILAC NEOSURE, ready-to-feed, with ARA and DHA[a]	69	3.77	0.022	0.012
ABBOTT NUTRITION, SIMILAC, ADVANCE, with iron, liquid concentrate, not reconstituted[a]	127	6.89	0.046	0.023
ABBOTT NUTRITION, SIMILAC, ADVANCE, with iron, powder, not reconstituted[a]	522	28.87	0.187	0.96
ABBOTT NUTRITION, SIMILAC, ALIMENTUM, ADVANCE, ready-to-feed, with ARA and DHA[a]	67	3.63	0.018	0.011
ABBOTT NUTRITION, SIMILAC, ISOMIL, ADVANCE with iron, powder, not reconstituted[a]	517	28.09	0.187	0.096
ABBOTT NUTRITION, SIMILAC, NEOSURE, powder, with ARA and DHA[a]	520	28.33	0.174	0.090
ABBOTT NUTRITION, SIMILAC, SENSITIVE (lactose free), liquid concentrate, with ARA & DHA[a]	128	6.95	0.031	0.020
ABBOTT NUTRITION, SIMILAC, SENSITIVE (lactose free), powder, with ARA and DHA[a]	520	28.14	0.130	0.079
ABBOTT NUTRITION, SIMILAC, SENSITIVE (lactose free), ready-to-feed, with ARA and DHA[a]	68	3.74	0.017	0.011
ABBOTT NUTRITION, SIMILAC, SPECIAL CARE, ADVANCE 24, with iron, ready-to-feed, with ARA and DHA[a]	71	3.26	0.015	0.009
MEAD JOHNSON, ENFAMIL, ENFACARE LIPIL, ready-to-feed, with ARA and DHA	71	3.80	0.024	0.012

Table 2: cont....

MEAD JOHNSON, ENFAMIL LIPIL, with iron, ready-to-feed, with ARA and DHA	63	3.5	0.019	0.009
MEAD JOHNSON, ENFAMIL, AR LIPIL, powder, with ARA and DHA	509	25.97	0.164	0.0087
MEAD JOHNSON, ENFAMIL, AR LIPIL, ready-to-feed, with ARA and DHA	68	3.49	0.023	0.012
MEAD JOHNSON, ENFAMIL, ENFACARE LIPIL, powder, with ARA and DHA	508	26.00	0.170	0.080
MEAD JOHNSON, ENFAMIL, LACTOFREE LIPIL, with iron, powder, with ARA and DHA	521	28.00	0.180	0.090
MEAD JOHNSON, ENFAMIL, LACTOFREE, LIPIL, with iron, liquid concentrate, not reconstituted, with ARA and DHA	132	7.11	0.040	0.019
MEAD JOHNSON, ENFAMIL, LIPIL, low iron, liquid concentrate, with ARA and DHA	131	7.16	0.041	0.019
MEAD JOHNSON, ENFAMIL, LIPIL, low iron, powder, with ARA and DHA	511	27.00	0.175	0.090
MEAD JOHNSON, ENFAMIL, LIPIL, low iron, ready to feed, with ARA and DHA	64	3.50	0.019	0.009
MEAD JOHNSON, ENFAMIL, LIPIL, ready-to-feed, with ARA and DHA	66	3.50	0.019	0.009
MEAD JOHNSON, ENFAMIL, LIPIL, with iron, liquid concentrate, with ARA and DHA	131	6.96	0.040	0.019
MEAD JOHNSON, ENFAMIL, LIPIL, with iron, powder, with ARA and DHA	511	27.00	0.168	0.089
MEAD JOHNSON, ENFAMIL, NUTRAMIGEN LIPIL, with iron, powder, not reconstituted, with ARA and DHA	494	26.00	0.170	0.080
MEAD JOHNSON, ENFAMIL, NUTRAMIGEN LIPIL, with iron, ready to feed, with ARA and DHA	66	3.50	0.019	0.009
MEAD JOHNSON, ENFAMIL, NUTRAMIGEN, LIPIL, with iron, liquid concentrate not reconstituted, with ARA and DHA	126	6.73	0.37	0.017
MEAD JOHNSON, ENFAMIL, PROSOBEE LIPIL, with iron, powder, not reconstituted, ARA and DHA	510	27.00	0.170	0.085
MEAD JOHNSON, ENFAMIL, PROSOBEE, LIPIL, liquid concentrate, not reconstituted, ARA and DHA	131	7.11	0.039	0.019
MEAD JOHNSON, NEXT STEP, PROSOBEE LIPIL, powder, with ARA and DHA	480	21.00	0.161	0.081
MEAD JOHNSON, NEXT STEP, PROSOBEE, LIPIL, ready to feed, with ARA and DHA	67	2.91	0.019	0.009
MEAD JOHNSON, PROSOBEE LIPIL, with iron, ready to feed, with ARA and DHA	64	3.50	0.019	0.009
NESTLE, GOOD START SOY, with DHA AND ARA, powder	503	25.60	0.175	0.090
NESTLE, GOOD START SOY, with DHA and ARA, liquid concentrate	132	6.67	0.040	0.019
NESTLE, GOOD START SOY, with DHA and ARA, ready-to-feed	64	3.28	0.019	0.009
NESTLE, GOOD START SUPREME, with iron, DHA and ARA, prepared from liquid concentrate	66	3.37	0.023	0.011
NESTLE, GOOD START SUPREME, with iron, DHA and ARA, ready-to-feed	66	3.37	0.023	0.011
PBM PRODUCTS, ULTRA BRIGHT BEGIN- NINGS, liquid concentrate, not reconstituted[b]	130	7.00	0.044	0.023
PBM PRODUCTS, ULTRA BRIGHT BEGINNINGS, powder[b]	524	28.00	0.181	0.103
PBM PRODUCTS, ULTRA BRIGHT BEGINNINGS, ready-to-feed[b]	63	3.50	0.022	0.013
PBM PRODUCTS, ULTRA BRIGHT BEGINNINGS, soy, liquid concentrate[b]	126	7.00	0.045	0.025
PBM PRODUCTS, ULTRA BRIGHT BEGINNINGS, soy, powder[b]	508	27.20	0.183	0.102
PBM PRODUCTS, ULTRA BRIGHT BEGINNINGS, soy, ready-to-feed[b]	63	3.50	0.022	0.013
PBM PRODUCTS, store brand, soy, liquid concentrate, not reconstituted[b]	126	7.00	0.045	0.022
PBM PRODUCTS, store brand, soy, powder[b]	508	27.20	0.183	0.091
PBM PRODUCTS, store brand, soy, ready-to-feed[b]	63	3.50	0.022	0.011
PBM PRODUCTS, store brand, liquid concentrate, not reconstituted[b]	130	7.00	0.044	0.022

PBM PRODUCTS, store brand, powder[b]	524	28.00	0.181	0.103
PBM PRODUCTS, store brand, ready-to-feed[b]	63	3.50	0.022	0.011

*From Baby foods group [64]
[a] Formerly ROSS; [b] Formerly WYETH-AYERST
AA: arachidonic acid; DHA: docosahexaenoic acid.

DIETARY SOURCES OF N-3 PUFAs

Plant-based sources of ALA are mainly the group of seeds and nuts such as canola, soybeans, walnuts, flaxseed and oils obtained from these foods, being flaxseed the richest (Table **3**), whereas fish is the primary food source of EPA and DHA. On the other hand, LA from the n-6 family is widespread in nature. It is contained in high amounts particularly in seeds and oils of most plants except for coconut, cocoa and palm [28,63,64] (Table **3**). Marine food sources have the highest content of EPA and DHA (Table **4**). Optimal diet-based options are salmon, blue crabs, tuna canned in water (white), tuna (bluefin), trout (rainbow), cod, caviar, catfish, carp, bluefish, anchovy, and mackerel (except kingfish and bass from the Gulf of Mexico due to high levels of environmental contaminants, see above). Although EPA and DHA are contained in seafood and all fish oil supplements, the ratio between those two differs: fish in general is higher in DHA and supplements have higher EPA [64,65].

Table 3: Energy, lipids and n-3 fatty acids content in selected plant-based food: nuts, seeds and oils.

Food name*	Content in 100 g of edible portion			
	Energy, kcal	Lipids, g	AL, g	ALA, g
NUTS				
Almonds, unroasted	575	49.42	12.055	0.006
Brazil nuts, dried, unblanched	656	66.43	20.543	0.017
Coconut meat, dried (desiccated), sweetened, flaked	456	27.99	0.222	0.000
Pine nuts, dried, *Pinus* spp.	673	68.37	33.150	0.112
Walnuts, black, dried, *Juglans nigra*	618	59.00	33.072	2.006
Pumpkin and squash seed kernels, dried	559	49.05	20.667	0.120
Sunflower seed kernels, oil roasted, with salt added	592	51.30	34.124	0.076
Amaranth, uncooked, *Amaranthus* spp.	371	7.02	2.736	0.042
Cornmeal, degermed, enriched, white	369	1.79	0.715	0.000
OILS AND MARGARINE				
Olive, salad or cooking	844	100.00	9.762	0.761
Canola	844	100.00	18.640	9.137
Soybean, salad or cooking	844	100.00	50.418	6.789
Commodity food, oil, vegetable, soybean, refined	844	100.00	50.118	6.537
Flaxseed	844	100.00	12.700	53.300
Walnut	844	100.00	52.900	10.400
Avocado	844	100.00	12.530	0.957
Margarine-type vegetable oil spread, 70% fat (soybean and partially hydrogenated soybean, stick)	628	70.22	17.280	2.073

*From groups: nuts and seeds, and fats and oils [64].
AL: linoleic acid; ALA: alpha-linolenic acid.

Table 4: Energy, lipids and n-3 fatty acid content in selected marine-derived sources.

Food name*	Content in 100 g of edible portion			
	Energy, kcal	Lipids, g	EPA, g	DHA, g
Anchovy, European, canned in oil, drained solids	210	9.71	0.763	1.292
Anchovy, European, raw. *Engraulis encrasicholus* (L.)	131	4.84	0.538	0.911
Bass, fresh water, mixed species, raw. *Percichthyidae and Centrarchidae*	114	3.69	0.238	0.357
Bluefish, raw. *Pomatomus saltatrix* (L.)	124	4.24	0.252	0.519

Table 4: cont….

Burbot, raw. *Lota lota* (L.)	90	0.81	0.070	0.096
Carp, raw. *Cyprinus carpio* (L.)	127	5.60	0.238	0.114
Catfish, channel, farmed, raw. *Ictalurus punctatus*	135	7.59	0.067	0.207
Catfish, channel, wild, raw. *Ictalurus punctatus*	95	2.82	0.130	0.234
Caviar, black and red, granular	252	17.90	2.741	3.800
Cisco, raw. *Coregonus artedi* Lesueur	98	1.91	0.095	0.257
Cod, Atlantic, canned, solids and liquid	105	0.86	0.004	0.153
Cod, Atlantic, cooked, dry heat	105	0.86	0.004	0.154
Cod, Atlantic, dried and salted	290	2.37	0.011	0.423
Cod, Atlantic, raw. *Gadus morhua* (L.)	82	0.67	0.064	0.120
Cod, Pacific, cooked, dry heat	105	0.81	0.103	0.173
Cod, Pacific, raw. *Gadus macrocephalus* Tilesius	82	0.63	0.080	0.135
Croaker, Atlantic, raw. *Micropogonias undulatus* (L.)	104	3.17	0.123	0.097
Crustaceans, crab, Alaskan king, cooked, moist heat	97	1.54	0.295	0.118
Crustaceans, crab, Alaskan king, imitation surimi	95	0.46	0.0	0.022
Crustaceans, crab, blue, canned	99	1.23	0.193	0.170
Crustaceans, crab, blue, cooked, moist heat	102	1.77	0.243	0.231
Crustaceans, crab, blue, crab cakes	155	7.52	0.227	0.216
Crustaceans, crab, blue, raw	87	1.08	0.170	0.150
Crustaceans, crab, dungeness, cooked	110	1.24	0.281	0.113
Crustaceans, crab, dungeness, raw	86	0.97	0.219	0.088
Fish portions and sticks, frozen, preheated	249	13.25	0.150	0.252
Grouper, mixed species, raw. *Epinephelus* spp.	92	1.02	0.027	0.220
Halibut, Atlantic & Pacific, raw. *Hippoglossus hippoglossus*	110	2.29	0.071	0.292
Maddock, raw. *Melanogrammus aeglefinus*	87	0.72	0.059	0.126
Mackerel, Atlantic, raw. *Scomber scombrus*	205	13.89	0.898	1.401
Mackerel, Pacific and jack, mixed species, raw. *Scomber* spp. and *Trachurus* spp.	158	7.89	0.509	0.932
Mackerel, king, raw. *Scombermorus cavalla* (Cuvier)	105	2.00	0.136	0.177
Mackerel, spanish, raw. *Scombermorus maculatus*	139	6.30	0.329	1.012
Salmon, Atlantic, farmed, raw. *Salmo salar* L.	208	13.42	0.862	1.104
Salmon, Atlantic, wild, raw. *Salmo salar* L.	142	6.34	0.321	1.115
Salmon, chinook, raw. *Oncorhynchus tshawytscha*	179	10.43	1.008	0.944
Salmon, chum, raw. *Oncorhynchus keta* (Walbaum)	120	3.77	0.233	0.394
Salmon, coho, farmed, raw. *Oncorhynchus kisutch*	160	7.67	0.385	0.821
Salmon, coho, wild, raw. *Oncorhynchus kisutch*	146	5.93	0.429	0.656
Salmon, pink, raw. *Oncorhynchus gorbuscha*	116	3.45	0.419	0.586
Salmon, sockeye, raw. *Oncorhynchus nerka*	168	8.56	0.519	0.653
Smelt, rainbow, raw. *Osmerus mordax*	97	2.42	0.275	0.418
Trout, rainbow, farmed, raw. *Salmo gairdneri* Richardson	138	5.40	0.260	0.668
Trout, rainbow, wild, raw. *Salmo gairdneri* Richardson	119	3.46	0.167	0.420
Tuna, fresh, bluefin, raw. *Thunnus thynnus* (L.)	144	4.90	0.283	0.890
Tuna, fresh, skipjack, raw. *Euthynnus pelamis* (L.)	103	1.01	0.071	0.185
Tuna, fresh, yellowfin, raw. *Thunnus albacares*	108	0.95	0.037	0.181
Tuna, light, canned in water, drained solids	116	0.82	0.047	0.223
Tuna, light, canned in oil, drained solids	198	8.21	0.027	0.101
Tuna, white, canned in water, drained solids	128	2.97	0.233	0.629
Tuna, white, canned in oil, drained solids	186	8.08	0.066	0.178

		Table 4: cont....		
USDA Commodity, salmon nuggets, breaded, frozen	212	11.72	0.055	0.441
Scamp. *Mycteroperca phenax***	--	1.52	0.008	0.106
Striped searobin. *Prionotus evolans***	--	1.71	0.012	0.124
Tuna, . *Thunnus albacares***	--	0.76	0.016	0.152
Hake, North Pacific. *Merluccius productus***	--	1.07	0.023	0.102
Red snapper, Carribbean. *Ocyurus chrysurus***	--	1.23	0.028	0.237
Snapper, yellowtail. *Ocyurus chrysurus***	--	0.92	0.002	0.238

*From finfish and shellfish group [64].
**Analyzed without skin [65].
EPA: eicosapentaenoic acid; DHA: docosahexaenoic acid. --: No data.

Nutritionists generally recommend a food-based approach for achieving an adequate intake. However, for individuals who are allergic or cannot eat fish, those who do not like fish, and for vegans, EPA and DHA intake may be problematic. A supplement (Table **5**) or fortified/enriched food may be alternatively advised such as poultry meat, eggs and dairy products [5,66]. FDA has ruled that intakes of up to 3 g/day of marine n-3 fatty acids are Generally Recognized As Safe (GRAS) for consumption in the diet, including specific consideration of the reported effects of n-3 fatty acids on glycemic control in patients with diabetes, bleeding tendencies, and LDL-cholesterol. Moreover, the FDA has also approved a qualified health claim for EPA and DHA n-3 fatty acids in dietary supplements [25], verifying that recommended sources are free of methylmercury, PCBs and dioxin-like contaminants. To avoid exceeding the safe limits of intake of these environmental pollutants but to achieve desired EPA and DHA intakes, consumption of a variety of fish is recommended. In addition, the patient must know that supplements may have side effects such as gastrointestinal disturbances/upset and nausea [25].

Table 5: Energy, EPA and DHA content in some commercial supplements (per softgel)

Name of product	Energy, kcal	EPA, mg	DHA, mg
ABC Vitamin Life	10	360	240
Max EPA	10	180	120
Herbal Remedies	5	200	150
Omega Max	10	360	240
Nordic Naturals (Ultimate Omega)	10	325	225
NutraOrigin	10	420	240
Expecta Lipil[a]	5	-	200
Life's DHA	5	-	200
Cardiostat	11	540	170
VitalOils 1000	11	250	750
Omega RX[b]	9	400	200
Salmon Oil 1000[b]	10	180	120
Omega III Salmon Oil[b]	10.6	180	120
Epacure omega 3[b,c]	--	400	200
Pulse Omega 3[b]	5.3	107 EPA plus DHA	

EPA: eicosapentaenoic acid; DHA: docosahexaenoic acid.

[a] Expecta Lipil DHA supplement is for pregnant and breastfeeding mothers.

[b] Available in Mexico.

[c] Data not available.

- Not contained.

Modified from Kris-Etherton *et al.*[5]

In conclusion, the strongest evidence regarding n-3 LC-PUFAs benefits have been reported on the reduction of risk of CHD or cardiac death, together with antiarrhythmic effects with an average EPA and DHA intake of either fish consumption (two portions of fish/shellfish or one portion of fatty fish per week) or from 250 to 500 mg of EPA and DHA per day. Likewise, for pregnant, lactating, and childbearing women, as well as for infants, the expert consensus based on systematic literature reviews supports the recommendation of an average DHA intake of at least 200 mg/day for woman during those physiological stages. For infants, breastfeeding is strongly endorsed; therefore, mothers must have a balanced diet including preferably food-based sources. On the other hand, for cognitive decline or Alzheimer's disease and affective disorders, there is limited and controversial evidence to support the beneficial effects of EPA and DHA. Further research is needed to establish conclusions.

REFERENCES

[1]   Innis, S. M. Omega-3 fatty acids and neurodevelopment to 2 years of age: do we know enough for dietary recommendations?. *J. Pediatr.* Gastroenterol. *Nutr.,* **2009**, *48*, S16-S24.

[2]   Kelley, D. S. Modulation of human immune and inflammatory responses by dietary fatty acids. *Nutrition,* **2001**, *17*, 669-673.

[3]   Muskiet, F. A. J.; Fokkema, M. R.; Schaafsma, A.; Boersma, E.R.; Crawford, M. A. Is docosahexaenoic acid (DHA) essential? Lessons from DHA status regulation, our ancient diet, epidemiology and randomized controlled trials. *J. Nutr.,* **2004**, *134*, 183-186.

[4]   Institute of Medicine. *Dietary reference intakes for energy, carbohydrate, fiber, fat, fatty acids, cholesterol, protein, and amino acids (macronutrients),* National Academy Press, Washington, D.C., **2005**.

[5]   Kris-Etherton, P. M.; Grieger, J. A.; Etherton, T. D. Dietary reference intakes for DHA and EPA. *Prostaglandins Leukot. Essent. Fatty Acids,* **2009**, *81*, 99-104.

[6]   Harris, W.S.; Mozaffarian, D,; Lefevre, M.; Toner, C. D.; Colombo, J.; Cunnane, S. C.; Holden, J. M.; Klurfeld, D. M.; Morris, M. C.; Whelan, J. Towards establishing dietary reference intakes or eicosapentaenoic and docosahexaenoic acids. *J. Nutr.,* **2009**, *139*, 804S-819S.

[7]   Mozaffarian, D.; Rimm, E. B. Fish intake, contaminants, and human health: evaluating the risks and the benefits. *JAMA,* **2006**, *296*, 1885-1899.

[8]   Harris, W. S.; Kris-Etherton, P. M.; Harris, K. A. Intakes of long-chain omega 3 associated with reduced risks for death from coronary heart disease in healthy adults. *Curr. Atheroscler. Rep.,* **2008**, *10*, 503-509.

[9]   Burr, M. L.; Ashfield-Watt, P. A.; Dunstan, F. D.; Fehily, A. M.; Breay, P.; Ashton, T.; Zotos, P. C.; Haboubi, N. A.; Elwood, P. C. Lack of benefit of dietary advice to men with angina: results of a controlled trial. *Eur. J. Clin. Nutr.,* **2003**, *57*, 193-200.

[10]   Anderson, R. N.; Smith, L. B. National Vital Statistics Reports: deaths: leading causes for 2002, Division of Vital Statistics, Centers for Disease Control and Prevention. http://www.cdc/nchs/data/nvsr/nvsr53/nvsr53_17.pdf. (accessed February 9, **2010**).

[11]   Gruppo italiano per lo studio della sopravvivenza nell'infarto miocardico. Dietary supplementation with n-3 polyunsaturated fatty acids and vitamin D after myocardial infarction: results of the GISSI-Prevenzione trial. *Lancet,* **1999**, *354*, 447-55.

[12]   Leaf, A.; Albert, C. M.; Josephson, M.; Steinhaus, D.; Kluger, J.; Kang, J. X.; Cox, B.; Zhang, H.; Schoenfeld, D. Prevention of fatal arrhythmias in high-risk subjects by fish oil n-3 fatty acid intake. *Circulation,* **2005**, *112*, 2762-2768.

[13]   Brouwer, I. A.; Zock, P. L.; Camm, A. J.; Böcker, D.; Hauer, R. N.; Wever, E. F.; Dullemeijer, C.; Ronden, J. E.; Katan, M. B.; Lubinski, A.; Buschler, H.; Schouten, E. G. Effect of fish oil on ventricular tachyarrhythmia and death in patients with implantable cardioverter defibrillators: the Study on Omega-3 Fatty Acids (SOFA) and ventricular arrhythmia randomized trial. *JAMA,* **2006**, *295*, 2613-2619.

[14]   FAO/WHO. Diet, Nutrition, and the Prevention of Chronic Diseases. Technical Report 916, World Health Organization, Geneva, Switzerland, 2003. http://apps.who.int/bookorders/anglais/detart1.jsp?sesslan=1&codlan=1&codcol=10&codcch=916 (accessed January 16, **2010**).

[15]   National Heart Foundation of Australia. Fish, fish oils, n-3 polyunsaturated fatty acids and cardiovascular health, 2008. http://www.heartfoundation.org.au/SiteCollectionDocuments/HW%20FS%20FishOil%20RevEv.pdf (accessed January 26, **2010**).

[16]   International Society for the Study of Fatty Acids and Lipids (ISSFAL). ISSFAL Policy Statement 3. Recommendations for intake of polyunsaturated fatty acids in healthy adults, 2004. http://www.issfal.org.uk/ (accessed January 18, **2010**).

[17]   Institute of Medicine. *Seafood choices balancing benefits and risks,* National Academy Press: Washington, D. C., **2006**.

[18]   United Kingdom Scientific Advisory Committee on Nutrition. Advice on fish consumption: benefits and risks, 2004. http://cot.food.gov.uk/pdfs/fishreport2004full.pdf (accessed January 18, **2010**).

[19]   Superior Health Council of Belgium. Advisory report: Recommendations and claims made on omega-3 fatty acids (SHC 7945), 2004.   https://portal.health.fgov.be/pls/portal/docs/PAGE/INTERNET_PG/HOMEPAGE_MENU/ABOUTUS1_   MENU

/INSTITUTIONSAPPARENTEES1_MENU/HOGEGEZONDHEIDSRAAD1_MENU/ ADVIEZENENAANBEVELINGEN1 _MENU/ADVIEZENENAANBEVELINGEN1_DOCS/OMEGA-3%20ENGLISH.PDF (accessed January 27, **2010**).

[20]   Health Council of the Netherlands. Guidelines to a healthy diet 2006. The Hague: Health Council of the Netherlands, 2006; publication no. 2006/21E.

[21]   Australia and New Zealand National Health and Medical Research Council. Nutrient reference values for Australia and New Zealand including recommended dietary intakes, Reference no: N35, N36, N37, 2006. http://www.nhmrc.gov.au/publications/synopses/n35syn.htm (accessed January 19, **2010**).

[22]   Backer, G.; Ambrosioni, E.; Borch-Johnsen, K.; Brotons, C.; Cifkova, R.; Dallongeville, J.; Ebrahim, S.; Faergeman, O.; Graham, I.; Mancia, G.; Manger-Cats, V.; Orth-Gomér, K.; Perk, J.; Pyörälä, K.; Rodicio, J. L.; Sans, S.; Sansoy, V.; Sechtem, U.; Silber, S.; Thomsen, T.; Wood, D. Third Joint Task Force of European and Other Societies on Cardiovascular Disease Prevention in Clinical Practice. European guidelines on cardiovascular disease prevention in clinical practice. *Eur. Heart. J.*, **2003**, *24*, 1601-1610.

[23]   Kris-Etherton, P. M.; Innis, S. Position of the American Dietetic Association and Dietitians of Canada: dietary fatty acids. *J. Am. Diet. Assoc.,* **2007**, *107*, 1599-1611.

[24]   Bantle, J. P.; Wylie-Rosett, J.; Albright, A. L.; Apovian, C. M.; Clark, N. G.; Franz, M. J.; Hoogwerf, B. J.; Lichtenstein, A. H.; Mayer-Davis, E.; Mooradian, A. D.; Wheeler, M. L. Nutrition recommendations and interventions for diabetes: a position statement of the American Diabetes Association. *Diabetes Care*, **2008**, *31*, S61-S78.

[25]   Kris-Etherton, P. M.; Harris, W. S.; Appel, L. J. Fish consumption, fish oil, omega-3 fatty acids, and cardiovascular disease. *Circulation*, **2002**, *106*, 2747-2757.

[26]   US Department of Health and Human Services and US Department of Agriculture. Dietary Guidelines for Americans, 2005. Government Printing Office, Washington, D.C. http://www.health.gov/dietaryguidelines/dga2005/report/ (accessed January 18, **2010**).

[27]   Deckelbaum, R. J.; Leaf, A.; Mozaffarian, D.; Jacobson, T. A.; Harris, W. S.; Akabas, S. R. Conclusions and Recommendations from the symposium, beyond cholesterol: prevention and treatment of coronary heart disease with n-3 fatty acids. *Am. J. Clin. Nutr.,* **2008**, *87*, 2010S-2012S.

[28]   Simopoulos, A. P. The importance of the omega-6/omega-3 fatty acid ratio in cardiovascular disease and other chronic diseases. *Exp. Biol. Med.*, **2008**, *233*, 674-688.

[29]   Morris, M. C. The role of nutrition in Alzheimer's disease: epidemiological evidence. *Eur. J. Neurol.*, **2009**, *16* (suppl. 1), 1-7.

[30]   Cole, G. M.; Ma, Q. L.; Frautschy S. A. Omega-3 fatty acids and dementia. *Prostaglandins Leukot. Essent. Fatty Acids*, **2009**, *81*, 213-221.

[31]   Conquer, L. A.; Tierney, M. C.; Zecevic, J.; Bettger, W. J.; Fisher, R. H. Fatty acid analysis of blood plasma of patients with Alzheimer's disease, other types of dementia, and cognitive impairment. *Lipids*, **2000**, *35*, 1305-1312.

[32]   Heude, B.; Ducimetiere, P.; Berr, C. Cognitive decline and fatty acid composition of erythrocyte membranes—the EVA Study. *Am. J. Clin. Nutr.*, **2003**, *77*, 803-838.

[33]   Tully, A. M.; Roche, H. M.; Doyle, R.; Fallon, C.; Bruce, I.; Lawlor, B.; Coakley, D.; Gibney, M. J. Low serum cholesteryl ester-docosahexaenoic acid levels in Alzheimer's disease: a case-control study. *Br. J. Nutr.*, **2003**, *89*, 843-849.

[34]   Kalmijn, S.; Feskens, E.; Launer, L.; Kronhout, D. Polyunsaturated fatty acids, antioxidants and cognitive function in very old men. *Am. J. Epidemiol.*, **1997**, *145*, 33-41.

[35]   Kalmijn, S.; Launer, L. J.; Ott, A.; Witteman, A. J.; Hofman, A.; Breteler, M. M. Dietary fat intake and the risk of incident dementia in the Rotterdam Study. *Ann. Neurol.*, **1997**, *42*, 776-782.

[36]   Morris, M. C.; Evans, D. A., Bienias, J. L.; Tangney, C. C.; Bennett, D. A.; Wilson, R. S.; Aggarwal, N.; Schneider, J. Consumption of fish and n-3 fatty acids and risk of incident Alzheimer disease. *Arch. Neurol.*, **2003**, *60*, 940-946.

[37]   Kalmijn, S.; van Boxtel, M. P.; Ocke, O.; Verschuren W. M.; Kromhout, D.; Launer, L. J.; Dietary intake of fatty acids and fish in relation to cognitive performance at middle age. *Neurology*, **2004**, *62*, 275-280.

[38]   Huang, T. L.; Zandi, P. P.; Tucker, K. L.; Fitzpatrick, A. L.; Kuller, L. H.; Fried, L. P.; Burke G. L.; Carlson M. C.; Benefits of fatty fish in dementia risk are stronger for those without APOE epsilon4. *Neurology*, **2005**, *65*, 1409-1414.

[39]   Nurk, E.; Drevon, C. A.; Refsum, H.; Solvoll, K.; Voll, S. E.; Nygard, O.; Nygaard, H. A.; Engedal, K.; Tell, G. S.; Smith, A. D. Cognitive performance among the elderly and dietary fish intake: the Hordaland Health Study. *Am. J. Clin. Nutr.*, **2007**, *86*, 1470-1478.

[40]   van Gelder, B. M.; Tijhuis, S.; Kalmijn, S.; Kromhout, D. Fish consumption, n-3 and subsequent 5-y cognitive decline in elderly men: the Zutphen Elderly Study. *Am. J. Clin. Nutr.*, **2007**, *85*, 1142-1147.

[41]   Morris, M. C.; Evans, D. A.; Tangney, C. C.; Bienias, J. L.; Wilson, R. S. Fish consumption and cognitive decline with age in a large community study. *Arch. Neurol.*, **2005**, *62*, 1849-1853.

[42]   Barberger-Gateau, P.; Raffaitin, C.; Letenneur, L.; Berr, C.; Tzourio, C.; Dartigues, J. F.; Alperovitch, A. Dietary patterns and risk of dementia: The Three-City cohort study. *Neurology*, **2007**, *69*, 1921-1930.

[43] Devore, E. E.; Grodstein, F.; van Rooij, F. J. A.; Hofman, A.; Rosner, B.; Stampfer, M. J.; Witteman, J. C. M.; Breteler, M. M. B. Dietary intake of fish and omega-3 fatty acids in relation to long-term dementia risk. *Am. J. Clin. Nutr.*, **2009**, *90*, 170-176.

[44] Freund-Levi, Y.; Eriksdotter-Jönhagen, M.; Cederholm, T.; Basun, H.; Faxén-Irving, G.; Garlind A.; Vedin, I.; Vessby, B.; Wahlund, L. O.; Palmblad, J. ω-3 fatty acid treatment in 174 patients with mild to moderate Alzheimer disease: OmegAD study. *Arch. Neurol.*, **2006**, *63*, 1402-1408.

[45] Milte, C. M.; Sinn, N.; Howe, P. R. C. Polyunsaturated fatty acid status in attention deficit hyperactivity disorder, depression, and Alzheimer's disease: towards an omega-3 index for mental health?. *Nutr. Rev.*, **2009**, *67*, 573-590.

[46] International Society for the Study of Fatty Acids and Lipids (ISSFAL). ISSFAL Official Statement 5, 2009. α-Linolenic acid supplementation and conversion to n-3 long chain polyunsaturated fatty acids in humans. http://www.issfal.org.uk/index.php/pufa-recommendations-mainmenu-146 (accessed January 18, **2010**)

[47] McNamara, R. K. Evaluation of docosahexaenoic acid deficiency as a preventable risk factor for recurrent affective disorders: current status, future directions, and dietary recommendations. *Prostaglandins Leukot. Essent. Fatty Acids*, **2009**, *81*, 223-231.

[48] Cao, J.; Schwichtenberg, K. A.; Hanson, N. Q.; Tsai, M. Y. Incorporation and clearance of omega-3 fatty acids in erythrocyte membranes and plasma phospholipids. *Clin. Chem.*, **2006**, *52*, 2265-2272.

[49] Harris, W. S.; Pottala, J. V.; Sands, S. A.; Jones, P. G. Comparison of the effects of fish and fish-oil capsules on the n-3 fatty acids content of blood cells and plasma phospholipids. *Am. J. Clin. Nutr.*, **2007**, *86*, 1621-1625.

[50] Arteburn, L. M.; Hall, E. B.; Oken, H. Distribution, interconversion, and dose response of n-3 fatty acids in humans. *Am J. Clin. Nutr.*, **2006**, *83*, 1467S-1476S.

[51] Hibbeln, J. R.; Nieminen, L. R.; Blasbalg, T. L.; Riggs, J. A.; Lands, W. E. Healthy intakes of n-3 and n-6 fatty acids: estimations considering worldwide diversity. *Am. J. Clin. Nutr.*, **2006**, *83*, 1483S-1493S.

[52] Lapillonne, A.; Jensen, C. L. Reevaluation of the DHA requirement for the premature infant. *Prostaglandins Leukot. Essent. Fatty Acids*, **2009**, *81*, 143-150.

[53] Burdge, G. C.; Wootton, S. A. Conversion of alpha-linolenic acid to eicosapentaenoic, docosahexaenoic and docosahexaenoic acids in young women. *Br. J. Nutr.*, **2002**, *88*, 411-420.

[54] Brenna, J. T.; Varamini, B.; Jensen, R. G.; Diersen-Schade, D. A.; Boettcher, J. A.; Arterburn, L. M. Docosahexaenoic and arachidonic acid concentrations in human breast milk worldwide. *Am. J. Clin. Nutr.*, **2007**, *85*, 1457-1464.

[55] Clandinin, M. T.; Chappell, J. E:; Heim, T.; Swyer, P. R.; Chance, G. W. Fatty acid utilization in perinatal de novo synthesis of tissues. *Early Hum. Dev.*, **1981**, *5*, 355-366.

[56] Smithers, L. G.; Gibson, R. A.; McPhee, A.; Makrides M. Higher dose of docosahexaenoic acid in the neonatal period improves visual acuity of preterm infants: results of a randomized controlled trial. *Am. J. Clin. Nutr.*, **2008**, *88*, 1049-1056.

[57] Koletzko, B.; Cetin I.; Brenna, T. J. for Perinatal Lipid Intake Working Group. Dietary fat intakes for pregnant and lactating women. *Br. J. Nutr.*, **2007**, *98*, 873-877.

[58] Koletzko, B.; Lien, E.; Agostoni, C.; Böhles, H.; Campoy, C.; Cetin, I.; Decsi, T.; Dudenhausen, J. W.; Dupont, C.; Forsyth, S.; Hoesli, I.; Holzgreve, W.; Lapillonne, A.; Putet, G.; Secher, N. J.; Symonds, M.; Szajewska, H.; Willatts, P.; Uauy, R. The roles of long-chain polyunsaturated fatty acids in pregnancy, lactation and infancy: review of current knowledge and consensus recommendations. *J. Perinat. Med.*, **2008**, *36*, 5-14.

[59] Codex Alimentarius Commission. Report of the 28th Session of the CODEX Committee of Nutrition and Foods for Special Dietary Uses. Codex Alimentarius Commission, 30 October–3 November, **2006**.

[60] Koletzko, B.; Baker, S.; Cleghorn, G.; Neto, U. F.; Gopalan, S.; Hernell, O.; Hock, Q. S.; Jirapinyo, P.; Lonnerdal, B.; Pencharz, P.; Pzyrembel, H.; Ramirez-Mayans, J.; Shamir, R.; Turck, D.; Yamashiro, Y.; Zong-Yi, D. Global standard for the composition of infant formula: recommendations of an ESPGHAN coordinated international expert group. *J. Pediatr. Gastroenterol. Nutr.*, **2005**, *41*, 584-599.

[61] The Commission of the European Communities. Commission Directive 2006/141/S of 22 December 2006 on infant formulae and amending Directive 1999/21/EC. *Official Journal of the European Union*, 30.12.2006, L 401/1-L 401/33. http://eur-lex.europa.eu/LexUriServ/LexUriServ.do?uri=OJ:L:2006:401:0001:0033:EN:PDF (accessed January 31, **2010**).

[62] International Society for the Study of Fatty Acids and Lipids (ISSFAL). Recommendations of the Second Statement on Dietary Fats in Infant Nutrition, 2008. http://www.issfal.org.uk/index.php/pufa-recommendations-mainmenu-146. (accessed January 16, **2010**).

[63] Gebauer, S. K.; Psota, T. L.; Harris, W. S.; Kris-Etherton, P. M. n-3 fatty acids dietary recommendations and food sources to achieve essentiality and cardiovascular benefits. *Am. J. Clin. Nutr.*, **2006**, *83*, 1526S-1535S.

[64] USDA National nutrient database for standard reference, release 22 (SR 22). Last Modified: 11/20/2009 [database on internet]. Washington, D.C.: USDA, Agricultural Research Service. http://www.ars.usda.gov/Services/docs.htm?docid=8964 (accessed January 16, **2010**).

[65]   Castro-González, M. I.; Ojeda, V. A.; Montaño, B.S.; Ledezma, C. E.; Pérez-Gil, R. F. Evaluation of the n-3 fatty acids from 18 |Mexican marine-fish species as functional foods. *Arch. Latinoam. Nutr.*, **2007**, *57*, 85-93.

[66]   Givens, D.I.; Gibbs, R. A. Current intakes of EPA and DHA in European populations and the potential of animal-derived foods to increase them. *Proc. Nutr. Soc.*, **2008**, *67*, 273-280.

SUBJECT INDEX

A

Acetylcholine 29, 31, 36, 37
Acetyl cholinesterase 24
Acyl-CoA 2, 5, 6, 42
Acetyl-CoA carboxylase (ACC) 7, 8
Acyl-CoA oxidase 5
Adhesion molecules 5, 8, 31, 41
Adipogenic genes 6
Adipokines 40, 51
Adiponectin 6, 39, 41, 51, 52
Adipose tissue 4, 6, 7, 12, 13, 17, 40-42, 46, 48, 51
Allergy 66-70, 73, 74
Alzheimer 24-26, 28, 30, 31, 81, 88
Anchovy 85
Angiotensin 40, 62
Apolipoprotein 24, 41
Asthma 1, 66, 68-72, 74
Atherosclerosis 4, 25-28, 59, 61, 80
Atopic dermatitis 68, 69, 73
Atopic eczema 70, 73
Atopy 68, 69

B

Beta-amyloid (Aβ) 23, 24, 29, 31
Bipolar disorder 60, 81
Bluefish 86
Brain 1, 3, 7, 11, 17-19, 23, 24, 29-31, 51, 60-62, 78, 82, 83

C

Canola 70, 85
Carbohydrate response-element binding protein (ChREBP) 1, 4, 5, 7, 8
Carbohydrate-responsive element (ChoRE) 5, 7, 8
Cardiovascular diseases (CVDs) 31, 58-62, 81
Carnitine *O*-palmitoyltransferase 5
Carp 85, 86
Caviar 85, 86
Central nervous system (CNS) 11, 12, 16-18, 81
Cerebrosides 18
Chemokines 5, 8
Citrate lyase 7
Cleavage-activating protein (SCAP) 7
Cod 11, 15, 85, 86
Cognitive function 18, 23, 25, 28-30, 32, 33
Cognitive impairment 23-25, 27, 29, 32, 33
Coronary heart disease (CHD) 62, 78-82, 87
Crabs 85
Cyclooxygenase 2, 8, 41, 51, 67, 69
Cytokines 5, 8, 31, 40, 41, 43, 51, 61, 68, 69, 71

www.ingramcontent.com/pod-product-compliance
Lightning Source LLC
Chambersburg PA
CBHW041720210326
41598CB00007B/729